THE COMPLETE
ENCYCLOPEDIA OF
RABBITS & RODENTS

THE COMPLETE
ENCYCLOPEDIA OF
RABBITS & RODENTS

Comprehensive information on
hamsters, mice, rats, gerbils, and guinea pigs

Also including less well-known pets, such as ferrets and chinchillas

ESTHER VERHOEF-VERHALLEN

REBO
PUBLISHERS

© 1997 Rebo International b.v., Lisse, The Netherlands

Text: Esther Verhoef-Verhallen
Translation: Stephen Challacombe
Cover design: Minkowsky Graphics, Enkhuizen, The Netherlands
Co-ordination, editing and production: TextCase, Groningen, The Netherlands
Typesetting and layout: Hof&Land Typografie, Maarssen, The Netherlands

ISBN 90 366 1596 8

Contents

Foreword

This encyclopaedia enables you to find out much about all manner of popular rodents such as guinea-pigs (or cavies), chinchillas, fancy rats, pet mice, degus, squirrels, and various types of hamster and gerbil. In addition, a substantial part of this encyclopaedia deals with some sixty different breeds of rabbit. For each type of animal a description is given of the breed characteristics, information about care required, accommodation, and breeding. This book is unique in its treatment of the subject, with many photographs to accompany this information.

It is a book to consult if you wish to find out whether a particular pet you consider is appropriate for you in terms of its behaviour and care requirements. If you are merely curious about the many breeds, colours, and varieties of coat that there are within the world of rodents and rabbits, this book will feed that curiosity. Perhaps you already have one or more of these creatures and wish to know more about their breeding or history. For each of these cases, this book is an excellent choice. This encyclopaedia is not intended though as a handbook for breeders or judges. Given the sheer number of types of animal and breeds dealt with, this book cannot be totally comprehensive. It has not been possible to give precise breed standards for each type covered because these can vary from country to country. These are in any event not set in concrete, and are adapted to developments and changing circumstances. In order to give you the reader the best possible oversight of the tremendous variety available in the world, average sizes, weights, colours, and varieties have been given together with an indication of where great differences exist between countries or where certain breeds or varieties are not yet recognised everywhere. If there is a particular animal or breed that you are considering, it is possible to obtain the national standard from the appropriate national organisation.

I had the enormous pleasure of actually seeing most of the animals illustrated in this book in real life. Contact with these animals and especially with their breeders was a marvellous experience. I am tremendously grateful for all the four-legged models who posed for my camera. It is them after all that this entire book is about. It would have been impossible to photograph them without the eager co-operation of their breeders and owners. These are people who devote all their free time, with a smile, to caring for, and being with animals and protecting the breeds. There are also many who worked in the background on the content of this book whom I wish to thank wholeheartedly for their effort and enthusiasm.

I sincerely hope that this encyclopaedia, which gave me such pleasure to write, will be a source of inspiration to others.

Esther J. J. Verhoef-Verhallen

Left: Degu

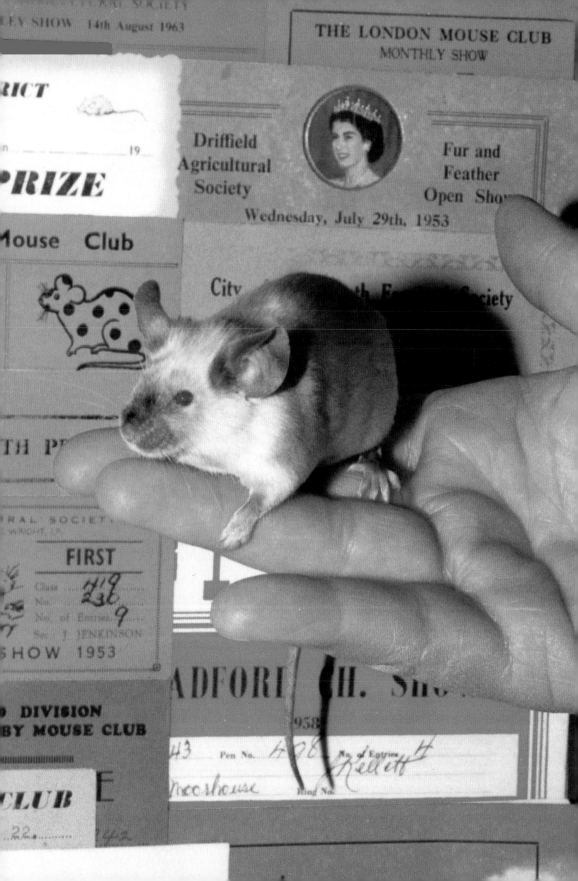

1. Introduction

Rodents

Most important characteristics

Incisor teeth

Rodents or the order of animals known as Rodentia are the largest group of mammals. The more than 400 genera and about 2,000 species of rodents make up over half of all mammalian species in the world. Rodents are to be found in widely varying parts of the world where they have adapted to extremes of habitat and climate.

It is a characteristic for the rodent to have large incising teeth. These continue to grow, even in the adult animal. Rodents therefore have to gnaw: this is how they wear their teeth at about the same rate as they grow. However, things can sometimes go amiss. If a tooth is broken, or the teeth do not develop evenly, so that the teeth no longer fit together correctly, the lack of resistance for the opposing tooth means there is insufficient wear so that the tooth continues to grow unrestricted. This causes the incisors to close the mouth, making it difficult or even impossible for the animal to eat and drink. In the wild, such an unfortunate animal would die but in the case of a pet, a vet can reduce the length of the teeth. The teeth will continue to grow, however, so the owner of such an animal will need to have the teeth dealt with several times each year. The teeth must fit together: that is a matter of life or death for a rodent.

Prolific reproduction

With a few notable exceptions, most rodents do not grow very old. Many of the smaller rodents only live a year or two, three at the most – and at that age they are positively ancient. The majority of wild rodents do not reach this sort of age. Small rodents are easy prey for all kinds and sizes of preying creatures.
Birds of prey, canine and feline pre-dators, but also many reptiles feed on small rodents. Rodents of very small size find the world a very dangerous place – they can barely defend themselves.
They have sharp teeth but these are no real protection against much larger enemies. Rodents do have one survival "weapon" against the heartlessness of nature in addition to their innate caution and intelligence and that is their sheer numbers.

Most small rodents can reproduce at a very young age. Some hamsters have been known to become sexually active at five weeks old. The gestation period for most small rodents

Left: This Siamese coloured mouse has won many distinctions.

Young three-coloured rex guinea-pig.

Most small rodents reproduce prolifically.

is also very short. Female mice, for example, can become pregnant again immediately after they have given birth and can therefore carry a second brood whilst the first are still suckling. Small rodents also produce fairly large litters. Domestic rats and mice lead the way with litters of ten to twenty being possible, although the average is somewhat lower. The young grow remarkably quickly and develop rapidly.

This prolific reproduction is a necessity for small rodents: if they could not reproduce so copiously, many of their species would have died out long ago.

An original agouti-coloured fancy rat.

Males fancy rats are far larger than the females.

A two weeks old white mouse.

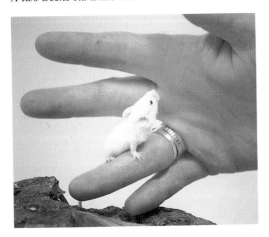

Nocturnal activity

The majority of small rodents are most active at night, the late afternoon, or early morning. During the day they mostly sleep and do not like being disturbed.

The reason for this nocturnal preference is that the majority of predators are active during the day, when they can see their prey more readily by daylight. Rodents that are kept as pets have no natural enemies, yet most of them keep to the natural life rhythms for their species.

A number of them, including hamsters, are therefore not particularly suitable as pets for small children. These animals are at their most active when the children go to bed. Their busy scrabbling can keep children awake at night if their cage is in the child's bedroom. The natural life rhythm of these pets is better suited to older children at

Syrian hamster: yellow on top, golden underneath.

school, or people who work during the day. Not all rodents are nocturnal though – some types of squirrel and also animals such as guinea-pigs are active during the day.

Other rodents have an irregular life pattern. They sleep for a while, are then active for an hour or two before having another nap. The dwarf Russian hamster is a good example of this.

Domestication

From rodent to pet

Many different rodents are featured in this encyclopaedia but the emphasis is on those that are widely kept as pets. The term pet can of course be interpreted widely. Most people understand a pet to be an animal that is sufficiently tame not to bite if it is picked up and that perhaps will permit lots of stroking and cuddling. These people choose a guinea-pig, a hamster, gerbil, or a mouse.

These animals are good choices, since they are sociable; they are domesticated. That means that that have been handled by humans and become pets. They have become so dependent on humans for their daily care that they could barely survive in the wild. These animals often have a wide variety of different types and colours of coat.

Hairless mouse.

This is another result of domestication. Think how the long-haired rodents would cope if they had to fend for themselves or how hairless mice would be unable to survive winters in the wild.

The bright colours and patterns which have been created in captivity would also reduce such an animal's chances of survival because they would stand out against the bushes and grass to the advantage of predators. Domestication can change more than the physical appearance of such animals.

Even their reproduction, social behaviour, and feeding can be slowly modified from the original characteristics of the species. This domestication is not harmful for the animal provided that it remains captive – and it applies to other animals as well as rodents. The many breeds of domesticated dog are far removed from their primitive forebear, the wolf. Milking cows and beef cattle too are

Degu

Fat-tailed gerbil

Negev gerbil

no longer capable of living without the care of humans. We must therefore take good care of our pet because it must never, ever be released into the wild.

Non-domesticated rodents as a pet

There are also rodents that have not been domesticated but which are kept as pets. This includes animals such as the Siberian chipmunk and the African spiny mouse. Various animals in this group will find their way, as did the Syrian hamster, the guinea pig, and the house mouse, to domesticity. Some of these animals exhibit remarkable differences from their native brethren in the wild.

They can be more susceptible (but sometimes less) to stress than their compatriots for example. Not all of these animals will, however, become fully domesticated. Some

European dwarf mouse

Siberian chipmunk

mimics the wild habitat as closely as possible and to watch the animal's natural behaviour.

Keeping non-domesticated rodents is very interesting but demands considerable insight and knowledge on the part of the keeper.

Never acquire one on impulse because inexperienced keepers are ill equipped to care for them properly without specialist guidance.

If you are interested in such an animal, get in touch with an association which has people who can advise you.

Acquiring a rodent

Future champion

Before acquiring a rodent you need to decide whether you are simply going to keep the animal as a pet or whether you want to show it and breed from it. This choice need not be made when you choose something exotic such as a Siberian chipmunk. These animals are not judged at shows.

If however you decide to keep a pet mouse, a fancy rat, golden hamster, or a guinea-pig (cavy) this choice is an important one. The standards for showing are very high and these should apply to the animals which are

of them adapt to a certain extent to their new situation by getting a little more used to humans.

A few may learn to take food from a human hand and even enjoy being stroked, but keeping this type of animal still requires very specific conditions and particular feeding requirements so that keeping them is really only a matter for experienced specialists. Keeping a non-domesticated rodent cannot therefore be compared with a mouse or guinea pig.

Enthusiasts for these exotic rodents are aware that they cannot expect the kind of affection that might be expected from the domesticated kinds. These animals will be wholly tamed, even if they are handled from a young age.

The true enthusiast does not feel it necessary to handle these animals. They find it far more exciting to create accommodation that

Gerbil

to form the foundation of your breeding stock. In such an event it is best to get advice from an association involved with showing and breeding the specific animals. Their members can ensure you choose suitable specimens.

It is a good idea to first visit a number of major shows to see as many animals as possible. Talk with as many breeders as you can and observe which types of animals always keep winning, You will gradually get a better idea and be far better prepared to proceed.

This encyclopaedia deals with the various colours and types of coat. It also addresses issues like the conformation (or body shape)

Young rex guinea-pigs

Because Syrian hamsters love to climb, a cage with horizontal bars is an ideal choice.

Standard-coloured chinchilla

American "crowned" black guinea-pig

of the animals. It may be that certain colours or varieties are not yet recognised and although you may have set your heart on these, it is wisest to start out with a colour choice or variety that is recognised, which will therefore be able to take part in shows. There are addresses of appropriate organizations at the back of the book.

The rodent as pet

If you are looking for a rodent simply to be a pet, then it is an easier task. If the animal has a spot or two too many or too few, or a conformation that is not according to the ideal for showing, there is no reason not to acquire the creature.

It could be that you find the very points which make the animal unsuitable for showing give it a special charm.

In this case you can purchase the animal from a pet shop but it is also possible to acquire such animals from breeders for they do not solely have perfect specimens. They may have less than ideal animals in a litter that they want to find a good home for. The big advantage when buying from a breeder is that you can learn more about the animal's background.

The parents will often be present and you can find out how the animal was brought up and what food it has been used to.
Try to avoid the more commercially-minded breeders because mice and rats are also bred on a large scale as food for snakes and other animals. It is obvious that breeding of healthy, sound animals is of no importance to such breeders: for them the quantity is far more important.

The breeding animals in these places produce many litters in a short space of time with little chance to recover. The mothers

A slightly older dwarf Russian hamster.

are increasingly unable to provide for their offspring and it is not surprising that animals from such breeding "factories" often bring their new owners nothing but grief because

Lilac-agouti multi-coloured guinea-pig with rex oat.

A breeding pair of guinea-pigs with Russian markings.

Such mice cages are widely used.

they have a weak constitution, leading to premature death.

Take the trouble to inform yourself about the previous history of the animals you have your eye on and only buy pets from a trustworthy source.

Animals must never be taken away from their mothers too early.

Pale gerbil

The purchase

When buying a rodent, especially if it is for a child, you must make sure that it is young, so that it can be tamed. It is possible to tame older rodents but this is more difficult. Do not buy animals that are too young though because when animals are taken away from their mother's too soon they can develop all kinds of behavioural problems and they are also often too weak and underdeveloped to cope without their mother.
If you are uncertain about such matters, take someone with you who does know, or buy only from a breeder with a good reputation. Take note of the length of the incisor teeth, for great long "tusks" occur quite frequently with rodents. In these cases, the incisors do not meet properly and hence do not wear correctly.

Since such teeth continue to grow, if they are not attended to, the teeth will make it impossible for the animal to eat. A vet or experienced breeder can reduce the length of the teeth but it will continue to be an annoying problem that will never be cured.

Degus are very inquisitive

Golden Syrian hamster

Mouse hamsters

Natural-coloured dwarf Campbelli hamster

It goes without saying that you should never breed from such an animal with a deformity. It is also important to check that the animal is healthy in other ways. The animal should be free of fleas and lice and its coat should be springy to the touch, without any bare patches or scabs.

Discharges from the nostrils are always suspicious and those with diarrhoea are best left where they are. Apart from this, an arched back, a coat that sheds hair, and swellings of the legs and feet are unfavourable.

The animal should be in good condition – not too thin, but also not too fat – and giving an impression of a lively zest for life.

Transporting your pet

You should always transport or carry your pet or pets as carefully as possible. Do not forget that rodents are so called because of their gnawing teeth which enable them to

Silver-sepia Syrian hamster

quickly chew their way out of paper or cardboard. Hamsters, fancy rats, degus, and mice are particularly adept at doing so. Guinea-pigs are far less of a problem in this respect. It is a good idea to buy a carrying case or to make one yourself. You may need to transport your animal regularly – to shows, for instance, or to the vet. A sound carrying case of wood or metal that has a top of sturdy but

Copper-coloured Syrian hamster with satin coat.

Chinchilla

Blue self colour mouse

Chinchilla in its sleeping refuge.

Degu

fine mesh is a good investment. There are also transparent plastic carrying cases available in the shops with a stout coloured or white lids that have openings and a hatch. These are not expensive and are ideal for carrying small rodents such as fancy rats, or mice, and hamsters.

Make sure that the animal will not become too cold or too hot during the journey. Most rodents are quite unable to withstand heat. Never leave a rodent in a car in the full sun and make sure there is sufficient ventilation. Never put animals that are intolerant of others together in a carrying case, even for a short time.

Fill the case with a good handful of hay so that the animal can hide in order to make itself feel safer. The hay also helps to cushion the animal if you have to brake suddenly in the car.

Breeding

Consider carefully

Before you start breeding animals, there are
various questions you must address. The
major one is: what is the purpose for which
you are breeding? Breeding to make money
rarely brings the hoped for results.

Housing, feeding, and taking care of rodents
all costs money and the price you may get for
selling any might just cover these costs if you
are really lucky. If you decide to house
the animals more cheaply, to change their
bedding less often, and to cut the cost of
their feed, the result will be seen clearly in
the animals' health and condition, and that
of their offspring.
Breeding must be regarded solely as a hobby.
Not everyone wants rodents as pets. It is
very difficult, for example, to find a good
home for a male mouse, because people find
their smell unpleasant. It may also be difficult to find homes for fancy rats. If you are a
member of an association or club, then you

*An adorable litter of Syrian hamsters, but what
is to happen to them when they get bigger?*

can exchange animals with each other but remember the animals will then have to meet the appropriate breed standard because there is little or no interest in animals of lesser standard among breeders and those who show rodents.

Before you breed a litter, you need to be certain of a home for any animals that are surplus to your own requirements. Find out whether the local pet shop is prepared to take any of them. Normally people will only take young animals. Older animals which cannot bred from any longer are of virtually no interest to anyone. Take those wonderful stories about the amazing prices that breeders can make for animals with a considerable amount of salt. People talk very readily of their successes but probably less easily about the countless times they have had to add more cages for the animals they cannot find homes for.

Other problems arise if you intend to breed animals that are not so well known as pets or which need very specific care. These sorts of animals cannot be given away to just anyone, because few people are aware of what is required to look after them. A responsible breeder will fully explain the care necessary and perhaps even offer help if things go wrong.

The aim of breeding

Breeding rodents is an enjoyable and fascinating hobby but the greatest satisfaction is generally achieved from breeding animals as closely as possible to the breed standard.

In this case you are not just busy reproducing animals – which anyone could do –

Siamese coloured mouse with satin coat.

Golden is the original colour of Syrian hamsters.

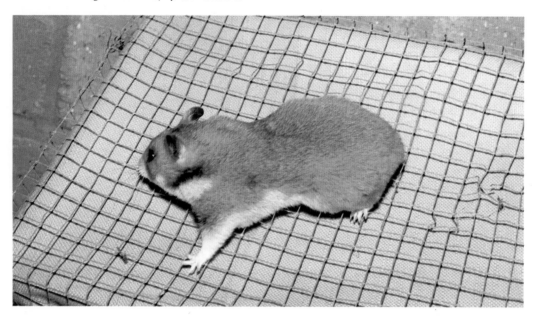

but seeking to improve the breed. In this case it is best to specialize in just the one or no more than two breeds.

To avoid inbreeding, you will need quite a few different breeding lines that are not or only distantly related to each other. If you raise several breeds, you will soon run out of space but also have too little time to look after all the animals properly. Money, as always, plays a role of course.

Breeding the most beautiful animals possible is a great challenge. The excitement to be had from realising that a litter consists of virtually perfectly formed or marked animals is unbeatable for the true enthusiast. Only experienced breeders who regularly attend shows know when they have such an animal.

They know the breed standards of their particular variety backwards and are aware of the points that judges particularly look for. As a newcomer to breeding, many of the characteristics of your animals will fly over your head at first, which is another good reason to join a club or association, and to show your own animals frequently.

In this way, you will learn from judges what they think of your animals and which points need special consideration.

You will also get to know other breeders and see a wider range of animals like yours,

which will give you a more solid grounding in the breed.

NB Because the large group of rodents have widely diverse means of reproducing, there are no general guidelines that can be given about breeding.

The necessary information is incorporated with the material about the individual species and varieties.

Orange flecked Syrian hamster.

Rabbits

A rabbit is not a rodent

The difference between rodents and rabbits

To the uninitiated, rabbits are considered as rodents, although in reality they belong to a different order of mammals. There are indeed certain similarities between rabbits and rodents – rabbits have incisor teeth, which continue to grow, – and in common with most rodents, their young are born without fur but quickly grow into adulthood.

In this respect cavia (guinea-pigs and other cavies) and chinchillas, which are still considered rodents by the experts, are notable exceptions. The gestation period is relatively long in these two groups of animals and their young are born with coats of hair. They do suckle their mothers but can quickly eat food and move about. The last word on these two exceptions has not yet been heard because international discussions are still going on about whether these animals are really rodents.

These discussions were concluded some time ago in the case of rabbits and hares, which are now assigned to the hare-like family of Lagomorpha. One of the reason for

Left: an exceptionally fine looking rabbit grey Papillon.

this is that in contrast with rodents, rabbits and hares have four incisors rather than two in their upper jaw. The difference between rabbits and hares in matters of reproduction is quite interesting.

Hares give birth to their young in a hidden place rather than in a burrow and in common with cavia and chinchillas, their young are fully formed, complete with hair, and can soon leave their nest to eat alongside their mother.

Rabbits have very short gestation periods, in common with rodents. Their young are born naked and helpless in an underground burrow where they are entirely reliant on their mother's milk until they are able to leave the burrow under the watchful eye of their mother after several weeks to take their first steps outside.
Cross-breeding between hares and rabbits is therefore not possible, because of these marked differences.

Acquiring a rabbit

Different breeds

There are many different breeds of rabbit throughout the world, that are also bred in a range of different colours and varieties. Many of these breeds have existed since the nineteenth century but others are of more recent origin.
Rabbit breeds can be roughly split into two groups: fancy and fur. Animals in both

A light rabbit grey Silver

Amber Mini Lop

Conforming to breed standards

Breed standards for rodents such as mice, hamsters, and fancy rats are fairly consistently judged in virtually every country that has shows for these animals. This means that a fancy rat in the USA is required to look the same as one in Britain or Germany.

The colour descriptions are also similar and the acceptance of new colours and coat varieties is usually agreed on an international basis, with the exception of Great Britain and the USA, which tend to run ahead of the other countries in terms of introductions of new colours and varieties. New colours and varieties almost always originate in these two countries.

There is far less uniformity of breed standards with rabbits than with these rodents. Certain colours may be recognized in one country but not another. An example of this is the Beveren which is virtually only ac-

Angora rabbits have long been bred for the wool from their coats.

groups may have been bred for economic reasons: their pelt, their meat, or both. The Angora provides angora wool and the Burgundy Yellow and New Zealand rabbits were historically true meat providing breeds. The fancy breeds include rabbits such as Holland Lop, Netherlands Dwarfs, Polish, but also the Belgian Hare.

These are breeds that were bred from their outset for an attractive appearance. It is also possible for fur and fancy to be combined: the Viennese Blue is a prime example. There are breeds that started life being bred for their fur but which became more of a fancy breed once the fur industry lost interest in their pelts.

Their coats were too thin, too mixed, or in some other way did not have the quality that the fur industry wanted. The Satin and Rex rabbits fall into this category.

cepted in blue and white in the country of origin and surrounding countries but can be seen at shows in Britain in a great variety of colours.

There are rabbits that are bred in almost every country where there are rabbit fanciers but with a different conformation and coat standard in each country. The breed can differ in weight, shape of the head, build, and length of the ears from country to country. The Viennese in Switzerland look quite different from those in Belgium.

Some differences occur right from the first existence of the breed. A specific breed may be developed along divergent lines by German, British, and French breeders to a different idea of perfection, developing quite apart national types.

Because it is interesting to know what various countries do in their breeding, this encyclopaedia has attempted to remain as international as space permits. This means that breeds or colours are included that may not be found at your local show.

If you have a serious interest in a breed or type that cannot be found in your part of the world, contact your national rabbit fancier's organization. Their members may well be able to tell you if the breed or variety is bred in your country or only elsewhere.

Large or small

The smallest breeds of rabbits are the Polish and Netherlands Dwarfs. In many countries, these dwarf rabbits weigh no more than 1kg (2lb 3oz) but may weigh 1 ½kg (3lb 4 ½oz). The heaviest breeds are the Giant Papillons, Flemish Giants, and French Lop with minimum weights laid down of 6–7kg (13lb 3 ½oz–15lb 4 ½oz) but they are usually much heavier.

These breeds are big as well as heavy. Flemish Giants, for instance, are about 80cm (31 ½in) long, measured from chest to the tail. There are medium-sized breeds that weigh around 2 ½–5kg (5lb 7oz–11lb). Breeds in this range include the Californian, Alaska, and Viennese rabbits. There are also breeds that fall between the medium and dwarf rabbits, such as Dutch, Tans, Thrian-

Grey French Lop beside a Giant Papillon and Flemish Giant – the heaviest breed of rabbit.

Mini Lop

up. A Flemish Giant, weighing 8kg (17 ½lb), is twice as heavy as the average cat and if they start thrashing about with their legs, it is no fun for a small child to look after them. Yet acquiring a larger breed is well worth while, for those who have sufficient space. The smaller breeds will usually find a home regardless of whether they are suitable for showing or not.

The main interest in large rabbits seems to be culinary. So if you have the space and do not mind whether a rabbit is large or small, choose a big one. These larger breeds are often affectionate, gentle-natured, and very cuddly.

tas, and Himalayan. These weigh less than 3kg (6lb 10 oz) .

It is not at all surprising that the most popular pet rabbits are the small and dwarf breeds. A small rabbit does not need such a big cage, eats less, and is generally cheaper to take care of. What is more, a small rabbit is easier – especially for children – to pick

Different characters

Disparate breeds of rabbit not only vary in looks, they can also have different characters. The smaller breeds tend to be livelier

Blue/grey Viennese

Black marked German Lop

Sallander rabbit

For so far as it is possible, this encyclopaedia describes the middle point or average character for a given breed.
You need to observe the animals carefully at the place that you intend to buy your rabbit.

Buck or doe?

Male rabbits are called "bucks" and female rabbits are known as "does." The principal difference in outward appearance of the male and female rabbits is in the shape of the head and proportions of their bodies. In the majority of breeds, the bucks have larger heads with fuller cheeks than the does. The bodies of the does are less muscular and often slightly longer.

Apart from these differences of physical appearance, the two genders also have character differences. Buck rabbits are usually more temperamental and much more readily

Blue Rhinelander

A litter of cross-bred rabbits

and more temperamental than the larger breeds, which in turn tend to be more even-tempered.

A lively nature manifests itself with a great deal of standing on the hind legs, being active, and vigilant, but unfortunately it can also lead to being fairly unmanageable. It is difficult to give a character assessment for each breed.

The nature of rabbits and their character defects are largely inherited. If a breeder of the predominantly phlegmatic New Zealand White breed chooses to only breed the most active and temperamental animals from his or her stock, it will not take long before all the animals are temperamental.

The opposite is also true. A breeder of a generally active breed can create a line of very calm rabbits through strict selection of the breeding stock. Breeders have their own preferences and you will notice that animals from the same breed may have widely differing natures.

aroused than does. The exception to this is when a doe is in season or has a young litter. During these periods, she tends to be less tolerant and is inclined to be very aggressive.

One or more rabbits?

Rabbits like to have the company of other rabbits and do not like being left on their own all day. If you will be away from home all day, or if the rabbit is to be kept in an outside hutch or run, it is better to have two rabbits.

It is often enough for rabbits if there is another one close by. It is therefore possible to keep two rabbits in separate cages, provided that they can see each other. Rabbits have preferences just like people and their characters can clash.

If you wish to keep rabbits in the same hutch or run, it is best to get two young does, preferable from the same litter. This reduces

the chance of any problems compared with putting two adult, or near adult rabbits together. Two bucks – the odd exception omitted – cannot be put in the same hutch or run, even if they are brothers and therefore know each other.

Bucks will almost certainly fight once they are sexually mature. It is also not a good idea to put a buck and a doe together because

Netherlands Dwarf with a Blanc de Hotot markings and an albino Polish rabbit.

A Netherlands Dwarf with Dutch markings should have brown eyes.

Silvers are inquisitive and temperamental rabbits.

New Zealand Red buck

you can then expect to have several litters every year. A doe prefers to withdraw when she has a litter but this is not possible if she shares a hutch, and can cause stress.

Buying a rabbit

Before you buy a rabbit, it is a good idea to think carefully about what you want from your new pet. If you are merely looking for a delightful pet you can either go to a breeder or pet shop where you can pick the rabbit that pleases you most. Happily this can be a crossbreed or a pedigree rabbit that is not suitable for showing.

For a pet, minor defects in breed standard are of no concern, provided the animal is sweet-natured and healthy. However, if you want to consider showing your pet or breeding from it, then the acquisition has to be more carefully considered. You will need to start with a rabbit that is as close as possible to the perfect breed standard. Such

Two Netherlands Dwarf rabbits.

animals are not to be found in pet shops but at shows and with established breeders. The best way to meet breeders is by visiting shows.

At the larger shows, you will find a number of breeders present with their animals in one place. A further advantage is that you can learn about any undesirable characteristics of the breed under consideration. Shows in

Netherlands Dwarf with Blanc de Hotot markings.

Angora rabbit

French Lop of three weeks.

your area are advertised in your local newspaper but contact the national rabbit fancier's organisation to find out when and where shows are being held in your area. Talk to as many breeders as possible, because they have a great deal of experience with the specific breed and if your are seriously interested, they will certainly help you. The majority of rabbits in this book will

An older Argenté de Champagne doe.

be found at international shows but it is quite possible that the colour you prefer is not bred or perhaps not even recognized in your country.

To avoid disappointment it is a good idea to contact the National Rabbit Council or other national organisation to check whether the breed and colour may be shown or not. Addresses are given in the back of this book.

Things to consider

Regardless of whether the rabbit is acquired from a breeder or a pet shop, there are a number of things that need to be carefully considered in order to make sure that your new pet is sound and healthy. The coat should be springy to the touch and free from fleas and lice and of course there should be no scabs or bald patches.

Discharges from the nose or eyes is always a suspicious sign and so is diarrhoea. Such

Young blue Giant Papillon.

specimens are best avoided. The animal should be in good condition – neither too thin or too fat – and create a lively impression. A fairly common problem with rabbits is for them to grow "tusks" or over-long teeth where the incisors do not fit properly and therefore do not wear correctly. Since the teeth continue to grow, this problem will eventually make it impossible for the animal to eat and a vet's intervention will be needed at frequent intervals.

This is a nuisance, which cannot be permanently cured. Check the teeth of the rabbit you consider buying to make sure that the teeth are healthy and that they fit. Finally, the age of a rabbit is a matter that has to be borne in mind.

A rabbit should never be taken away from its mother too early because such animals will be weaker and far more susceptible to illness through having failed to build up sufficient resistance and a greater risk that they will die young. A rabbit should remain with its mother for a minimum of six weeks – preferably somewhat longer – and it is only developed sufficiently to live independently by ten to twelve weeks. If you have little experience with rabbits, buy only from a reputable breeder or take someone who does know with you.

You may also choose to purchase a slightly older rabbit. Provided that the animal has had good experience of humans, it will quickly adapt to its new circumstances.

Broken black English Lop.

New Zealand White

A litter of young Thrianta rabbits.

large enough, a carrying box is better to be somewhat cramped. It is never intended that the rabbit should spend too long in the box and it is best that the animal should be supported by the sides of the box so that it will not be break a leg or receive any other injury if something untoward should happen while being transported.

Place a good bunch of hay in the box to help cushion the rabbit. In an emergency, a cat basket or plastic carrying box can be used. For very short trips, a really strong carton can be used but make some holes to ensure adequate ventilation and make sure the box is firmly closed because rabbits are outstanding jumpers. It should go without saying that you will not expose your rabbit to extremes of low or high temperatures and do not forget that the temperature in a car in the summer can climb steeply, giving a

Transporting rabbits

You can make a carrying box for your rabbit or buy ready-made ones. The box needs to be strong and have sufficient ventilation. In contrast with a cage, which can never be

Agouti-coloured Mini Lop.

Conveying a rabbit

This Rhineland rabbit checks the surroundings outside its carrying box.

rabbit major problems. At these times limit travel to necessity and try to keep changes of temperature to a minimum.

Caring for rabbits

Indoor or outdoor hutch

Rabbits are perfectly happy to be out of doors but do not forget that although rabbits are well-protected against the cold, they need shelter from draughts, the wet, strong sun, and hard frost. An outdoor hutch must be placed in a sheltered position, out of the wind and enjoying plenty of shade.

It is sensible to move the hutch into an unheated shed, garage, or outhouse during the cold winter months and also when the temperature shoots up in the summer. Never move a rabbit from a cold shed to a heated area or vice versa because there is a very real risk that it will die from the abrupt change in temperature.

If the rabbit is kept in a shed, make sure there is good ventilation and that daylight can penetrate to the hutch. Rabbits can also be kept indoors. The animals are quite clean and do not smell provided that their hutch is cleaned frequently.

It is quite possible to house train a rabbit so that it will use a cat litter tray and may wander freely about the house. Such a house must not contain any dangers for a rabbit such as electrical cables, poisonous house plants, and other perils for a rabbit must be out of its reach.

Housing a rabbit

There is a variety of rabbit hutches for sale that are mainly divided into indoor and out-door hutches. Indoors cages also come in a profusion of different models but these equally split into two main types: those with plastic floor trays and mesh sides and top, and those with plastic bottom and sides, with mesh in the lid.

The advantage of the first type is that they have good ventilation and the temperature does not shoot up in the summer but their disadvantage is that you will often find some of the wood shavings, feed, and hay falls out of the hutch. The major advantage of the closed hutch is that the surrounding area remains clean.
It is important with this type of hutch to make sure that it is never exposed to strong sunshine. Hutches for outdoors can also be purchased ready made from pet shops but any reasonably creative do-it-yourselfer can make a hutch from timber. The wood must be able to withstand moisture from both rain and urine.
A suitable outdoor hutch must always be raised off the ground and be well insulated. The roof of the hutch should extend some way over the hutch to prevent rain pene-

Rabbits can cope perfectly well in an outdoor hutch

The New Zealand White is very calm.

pressed pellets and mixed feeds. The disadvantage of mixed feeds is that the fussy rabbit will eat all the bits it likes and leave the rest, so that the rabbit does not get a balanced diet.

Pressed pellets do not have this problem because they contain virtually everything needed for a balanced diet for your rabbit. The amount that you should feed your rabbit depends upon the size of your rabbit and the amount of exercise it gets. Animals that get too little exercise will quickly get too fat partly because they use up too little energy but also because they eat too much out of boredom.

Rabbits that have sufficient room eat less and stay in better condition. There is no problem in leaving dry food for them for some time. Pregnant and suckling does generally need about double the amount of feed. Always give them as much as they will eat. If you always give your rabbit the same ready-prepared feed but want to change over, do this gradually, never from one day

Burgundy Yellow

tration. The best outdoor hutches have a separate sleeping section. Most ready-made hutches are far too small.

Rabbits that virtually never get the chance to stretch their legs on grass or to run around indoors, need a really spacious hutch. If you are unable or unwilling to let your rabbit run around on the loose indoors or the garden, a large outdoor run is ideal for the daytime. Such a run needs to have mesh underneath as well as on the top and sides since rabbits are such expert burrowers.

Never place such a run directly in the unshaded sun, and make sure that part of the run is covered to enable the rabbit to shelter from rain, sun, and wind.

Feeding

By contrast with rodents, which will happily eat an insect or worm, rabbits only eat plants, making them herbivores. There are numerous ready-made feeds for rabbits for sale, which can be loosely divided into

Rabbits only eat food derived from plants.

to the next. Rabbits are very sensitive to such changes.

In addition to dry pellets, every rabbit needs fibre in the form of dust- and fungus-free hay. This must always be present in the hutch in sufficient quantity that the rabbit can eat whatever it requires. Hay should not be put on the floor of the cage, where it will be fouled, but placed in a rack or manger. Water is also important to a rabbit in addition to hay. Rabbits that rarely eat fresh green matter need far more water than those that have ample fresh green vegetable produce. A rabbit drinks on average about a tenth of its body weight in water each day. A rabbit will knock over a water bowl inside its hutch or make it dirty so a drinking bottle fixed on the outside of the hutch is a better solution.

Rabbits – particularly young ones – cannot cope with food that is too rich in moisture. Lettuce, cabbage and brassica, beans, clover, fresh spring grass, and root vegetables such as turnips, beet, swede, and mangels are too heavy for the stomach and cause many rabbits to suffer from severe wind. Younger rabbits – up to about three to four months – can even die as a result. Only give this type of food to adult rabbits and then only in small portions. Food that is too cold because it has just come from the deep-freeze or refrigerator can cause similar problems. Make sure that food is at ambient temperature.

Suitable fresh vegetable matter for rabbits

- baby carrots
- endive
- kale
- pear
- apple
- dandelion
- shepherd's purse
- plantain
- nettle
- herbs (parsley, thyme)
- radish leaves
- twigs with leave from fruit trees and willows

Breeding

Consider it carefully first

Everyone who seriously wishes to breed rabbits has a number of matters that should be carefully considered before going ahead. The first consideration, is do you have sufficient space?
Bucks can virtually never be placed in the same hutch once sexually mature and keeping a doe in the same hutch is not recommended because they need to be able to bring up their young undisturbed.
In order to breed rabbits, you need a considerable number of hutches and sufficient space for them. Some people keep rabbits in well fenced enclosures but this is not practicable for proper selective breeding, because there can be no control over the reproduction of free-running animals.
In addition to sufficient space, you also need to be able to spare a great deal of time for the

Two young rabbits with Magpie markings.

Two young Deilenaar rabbits.

daily tasks of caring for your breeding stock. You will also need to have the necessary financial means because rabbits need good nutrition, straw and wood shavings as bedding and other things like drinking bottles, carrying cages and so on.

On top of this there will be veterinary costs such as inoculations. Then when you wish to go on holiday, you will need someone with the requisite skill and knowledge to look after the animals for you. A friendly neighbour or relative might be capable of looking after one or two rabbits but a complete breeding stock of rabbits is an entirely different matter. Do not forget either that the investment in the necessary hutches and runs for the rabbits will be not inconsiderable, even if you make them yourself.
There are people who think they can become rich from breeding rabbits or at least make a little bit of money on the side. If this is your objective, then my advice is forget it: breeding rabbits should really only be considered as a hobby.
Housing, feeding, and caring for rabbits is an expensive business. The sale of your rabbits – if everything goes well – may recover some of these costs but it will never be a worthwhile source of income.

Before breeding a litter of rabbits, you need to be sure that you can sell or find homes for those rabbits you do not want, or do not have the space to keep. Breeders who have established a reputation may well have waiting lists but people who have just started out can run up against problems when they attempt to sell their rabbits. The larger-sized rabbits can be particularly difficult to find homes for and these are the very breeds that usually have large litters. This then leaves only a choice between keeping the rabbits yourself or selling them to dealers in rabbits for slaughter. It is also difficult to find homes for older rabbits that can no longer be used for breeding.

The breeding objective

You may choose to breed a litter of two of rabbits because you will enjoy watching the young animals grow up but most people breed them for idealistic reasons. They are intent on breeding rabbits that are as close to the breed standard as possible. These people are therefore not engaged in reproducing rabbits – which anybody can do – but selective breeding. In order to breed selectively, you will need to have an understanding of genetics and a substantial knowledge of the preferred characteristics for the specific breed. Only experienced breeders, who regularly attend shows and are in contact with other breeders, know when they have a valuable animal. They know the standard for the breed and specific colour variety intimately and are aware of the aspects which are given most attention by show judges.

A breeder who is just starting out is less aware of all these things and this is a strong reason for joining the appropriate club or association in order to keep in touch with other breeders. Entering your own rabbits to shows regularly so that they can be judged is an essential part of breeding, in order to discover what you are doing right, and what needs further consideration.

How breeding is carried out

A doe is sufficiently developed at about nine months to be bred for the first time. Bucks are capable of breeding earlier than this – at about five months – but it is best to wait until a more fully developed to have a better chance to assess the buck's usefulness as a stud. The doe is always placed in the buck's hutch or enclosure to be covered and not the other way round because the doe can be aggressive in her own territory and the buck may be so busy inspecting the new territory that he has no time to cover the doe. If it is apparent that the doe wants nothing to do with being covered then try again a few days later, although there is usually no problem. The act of covering does not last long and it is best to observe while it is taking place. Once the covering has been accomplished, put the doe back in her own hutch.

Does prefer to give birth in a separate, dark place. The best solution for this is a wooden brood compartment in the run or attached to her own hutch. You can prepare this with a layer of straw but this is not necessary, because the doe will herself line the place she is to give birth with hay, straw, and tufts of hair from her own undercoat of fur. Make sure the lid of the hutch or box can be re-

New Zealand Red buck

moved easily so that you can assess the situation. A brood box allows you to clean the rest of the hutch without disturbing the doe or her offspring. It is important to give them as much undisturbed peace and quiet as possible. Many does are less responsive to their handler during their pregnancy and when they are rearing their young and may even be quite testy. This is a completely natural occurrence. The gestation period for rabbits varies from 28 to 31 days. Baby rabbits are born blind and virtually bald. Until the young start to take some solid feed – at about two to three weeks old – the doe needs additional nutrition. It helps the milk production if you give the doe a daily feed of lukewarm milk. Never give young animals too much fresh green fodder and then only in small amounts.

These young rabbits are susceptible to gastric disorders and can die from eating too much green matter. Make sure there is sufficient hay for the animals. The young rabbits must stay with their mother at least until six weeks old, even though they may be weaned. It is possible from this age to be put them in a hutch separate from the doe but they should not be taken to an entirely new environment before they are ten to twelve weeks old.

Inheritance in breeding

Mutations

What is a mutation?

Popular breeds of rabbits and rodents are usually bred in quite a range of different colours and varieties.

The ancestors of these animals did not, however, have this diversity. The white patches, longer coats, and unusual colour patterns have resulted from mutation. Mutations are spontaneously occurring changes from the inherited genetic imprint. They are a natural phenomenon that is as old as the earth and they occur in every type of animal, but also with humans, and plants. Mutations can manifest themselves in a variety of ways. Perhaps an animal is born without any coat, or with a different colour of coat, maybe without eyes, or eyes that are larger than usual.

Mutations can also affect an animal in ways that cannot readily be seen: for example altering their digestive system, or brain size. A mutation is a trick of nature although they can be created by exposure to radioactive substances, for example. Mutations can be inherited, so that a certain percentage of the offspring from a mutated animal will carry the same mutated characteristics as the father of mother.

The point of mutations

Mutations have a valuable function in nature – to help the species to survive. Imagine a colony of hamsters living on rocky terrain. The hamsters are blue-grey, like the colour of the rocks. If the hamsters were beige, they would be easy pickings for birds of prey and other predators. These would be able to spot the beige hamsters immediately against the blue-grey background.

Colour therefore provides protection and aids survival. If a mutant hamster is born with a noticeable beige-coloured coat, it would have little chance of survival. Before the animal was able to grow up and repro-

Left: Harlequins have eumelaninen and phaeomelaninen.

A bristly coat is genetically dominant with guinea-pigs.

A marked Mongolian gerbil

The subtle colours of the European harvest mouse ensure that it does not stand out from its surroundings.

In the wild, the clearly marked coat of this mouse would be spotted by predators.

Blue Angora rabbit: blue is a dilution of black.

duce, it would be eaten up. This mutation is therefore not beneficial and does not help the species to survive.

However, if the hamsters are forced by circumstances to move, for example, to an area of desert, or if their natural surroundings become changed into a sandy desert, the blue-grey colour is no longer protection. Under these circumstances, the mutant beige hamster is more likely to survive than the blue-grey variety and the beige hamster gets the chance to reproduce and because mutations affect the genotype (inherited genetic material) as well as the outward appearance (phenotype), the animal passes on the advantage of its different colour to its offspring, that in turn pass the advantage on to other, non-mutated hamsters.

Mutations are not always so beneficial and in many instances are even quite pointless. They are the toss of the dice from the hand of mother nature. Sometimes they are for the good but more frequently at that time and that particular place they serve no benefit for

the furtherance of the species. Mutations are only beneficial when they result in an advantage for the mutant that helps its species – albeit in devolved form – to continue to survive.

Mutations in modern breeding practice

Animals living in hutches or cages do not have enemies. If a mutated animal is born that is capable of living and able to reproduce, it will be able to breed, if allowed to do so. Such breeding is not usually a problem. People find something out of the ordinary rather interesting and most breeders like to keep an unusual animal.

Such a mutant animal can be used as the starting point for a new colour or coat variety. The animal is cross-bred with various different breeds. As many of the resulting offspring that exhibit the new characteristic, or are likely to carry the gene for it, are bred with each other as possible. Eventually a new breed is formed. Examples of established breeds that resulted from mutations are (with rabbits): Rex, Satin, and French Lop. Recent examples of mutations that have led to new breeding lines are the hairless fancy rats and those without tails that have occurred in the United States, where they have appeared at shows. There are also hairless rabbits and guinea-pigs but these have not yet been seen at shows.

These animals are solely bred and kept in laboratories. The stream of new characteristics is nowhere near coming to an end. Since mutations are quite natural, mutant animals continue to be born, so that we can expect a continued stream of new characteristics.

The description of colours and varieties in this book is a current day picture. An encyclopaedia setting out to cover the same range of material in 100 years time would

Hairless mouse

By combining the factor for tan with the factor for a white rump, mice with white rumps and tan have been created.

Amber and white Holland Lop.

need to be much larger in size. The Dwarf Russian Hamster currently only has three different colours, including the natural colour. This species has not long been kept as a pet but one can expect that in thirty years from now that there will probably be in the region of forty different colours plus different varieties of this species.

Selection

New breeds or different varieties can also be created by careful selection of the characteristics in breeding stock, as well as through mutation. By combining certain characteristics, a great deal of patience, and an understanding of genetics, breeders can create new breeds from existing ones.

Many breeds were created in this way. The Mini Lop, a small lop-eared breed resulted from cross-breeding lop-eared and dwarf rabbits. The Belgian Hare was also bred through careful selection. By continually selecting animals with the sleekest and thinnest build, breeders were eventually able to create a new breed that was very different to existing breeds of rabbit.

Guinea-pig breeders have cross satin-haired and rex guinea-pigs with the result that there are now guinea-pigs with both characteristics. Combining colour together also offers possibilities. The creation of new breeds and varieties is therefore not entirely left to the whim of nature but is largely determined by the breeders, who combined and recreate the various inherited genetic factors.

Colours

Melanin

The colour of a coat results from the reflection of light from pigments that are present in the hairs. The distribution and form of the pigment can vary in terms of the clusters of molecules, and whether thickly or thinly coated.

This pigmentation is melanin. All the colours that are found in rodents and

English Crown golden agouti guinea-pig.

Black English Crown guinea-pig.

Young brown-agouti Angora rabbit.

Rex rabbits have coats of a different texture.

Dalmatian markings on a Rex.

Albino Dwarf Polish

rabbits are caused by two basic type of melanin: phaeomelaninen leads to yellow or red hair and eumelaninen creates black. A mutation can cause the form or density of the melanin pigmentation to be changed so that a different colour of light is reflected, resulting in a different colour hair.

A well-know mutation of black melanin pigmentation leads to havana or chocolate-brown. Another mutation causes there to be less melanin in the hair so that the colour is optically diluted. This results in black appearing as blue, and chocolate as lilac.

Colours such as brown, lilac, and blue, result from the black melanin.

The melanin that gives rise to red can be equally modified, leading to colours such as cream.

Agouti

Many rodents and rabbits are only known in the wild with a colour ticking in their coat, known as natural or agouti.

The ticking is the result of a gene that inhibits the formation of the basic colour in some parts of the hair. This causes there to be bands of dark and light colour on each hair.

This Giant Chinchilla rabbit has patchy black ticks.

The darker colour is the actual colour of the animal but the lighter bands, resulting from the gene that causes the ticking, holds back that colour in places.

A golden-agouti animal has a basic black colour and if the gene for the agouti or wild colour was not present, the animal would be plain black. The same is true of animals with a brown, blue, or lilac tick: without the agouti gene, these animals would be plain brown, blue, or lilac.

Agouti occurs in many species and is not restricted to rodents and rabbits. If one thinks about cats, the agouti gene can result in four different types of tabby markings. These different "tabby" markings are not found with rabbits and rodents, instead, the ticking or agouti effect covers the entire body, with the exception of the belly and legs, with the bands of light and dark hair bands being evenly distributed.

Fat-tailed gerbil

Genetics

The benefit of genetics

It is essential for a breeder to have an understanding of genetics. Once a breeder has a grasp of genetics, he or she can reasonably predict what colour or type of coat will result from breeding of specific animals. Unfortunately, books on genetics are full of complicated terminology, codes, and diagrams, putting off many who wish to acquaint themselves. For higher level study of genetics, a certain amount of specialized terminology is required but things can be kept much simpler for those who wish to grasp the basic principles.

The genetics of coat colour and texture are not as involved as many think. For those who do not yet know how such genetic information as colour is passed on, the following section should be of help.

Basic genetics

Every cell in an animal's body has a number of pairs of chromosomes that bear the genetic blueprint.

Magpie rabbit

Degu

The number of chromosomes varies from one type of animal to the next, except for the seed cells of the males and egg cells of the females.

These cells do not contain any pairs of chromosomes, only single chromosomes. This is logical, since the combination of chromosomes from the male seed and female egg deliver new pairings of chromosomes. If seed and egg cells did contain pairs of chromosomes, then the animals born from these cells would have four pairs of chromosomes.

Nature ensures that the number always remain the same through splitting of the pairs in the egg and seed cells – known as meiosis – so that an offspring always inherits half of its characteristics from the father and half from the mother. The chromosomes carry the genes which bear the blueprint for the new life with information such as length of coat, colour, any markings, length of the legs, and shape of the ears.

The genes also carry inherited genetic information on less apparent characteristics too, like predisposition for a particular defect, the working of the digestive system, brains, and behavioural characteristics such as lively, or good-natured. The genes only contain information that is inherited from the parents – or inborn characteristics. The genes cannot pass on characteristics that the parent has acquired during its life. A rabbit which has had its hair shaved off will still produce offspring with normal hair, and a degu that has lost its tail in a mishap does not produce tailless progeny.

An animal's nature or character can be partially inherited. An animal may have inherited a tendency for aggression but when it is well cared for and does not encounter situation that cause aggression, this inherited trait may remain dormant.

Dominant and recessive genes

Influenced by the process of meiosis, the pairs of chromosomes split into two separate chromosomes. With natural fertilisation, a random male chromosome and equally random female chromosome come together. The pairs of chromosomes contain double genes, hence double information.

An animals can carry both the gene for blue hair and also for black. The genes for colour may of course be the same. The genes do not intermix. If that were the case, a white animal bred with a black one would produce grey offspring but this is not the case. This is because some genes are dominant and others are recessive.

Long-haired guinea-pig

The dominant genes will always "take control" over recessive genes and determine the outcome, resulting in the outward appearance characteristic of the parent from which the dominant gene was inherited. Recessive genes only make their presence felt if there is no dominant gene to repress them. If an animal acquires a recessive characteristic – say long hair – it cannot have

Young gerbils

Fancy rats

Gerbils

Tame rat

Syrian hamster

inherited a gene for short hair or it would have had short hair, since this is a dominant gene.

Genes are virtually always dominant or recessive in comparison to another gene. An exception to this is the melanin pigmentation characteristics in which both main types can occur alongside each other.

Himalayan markings are recessive.

Long-haired white mouse with satin gene.

A cross between an animal with melanin that results in black, blue, brown, or lilac with one that has melanin that produces red can result in a combination of colours. An example of this is with the Harlequin rabbit that has both black and red in its coat.

Mice with tortoiseshell markings and guinea-pigs with brindle and tortoiseshell coats have a combination of both types of melanin.

Examples

A full coloured coat is dominant over partial albino coats, which can be seen with part albino rabbits. The recessive gene which is responsible for part albino markings prevents dark pigmentation being formed on the body.

The underlying colour of the animal's coat is only found on extremities such as the legs,

The satin coat gives a metallic sheen to this dove grey and tan mouse

The satin coat gives a metallic sheen to this dove grey and tan mouse

Since albinos do not carry the gene for entirely-coloured coats (the dominant gene is always evidenced in the appearance), two albino rabbits cannot produce offspring with full coloured coats.

The special thing about recessive genes is that the characteristics which they carry the blueprint of can come to light although there is no certainty of it. The way in which genes come together is entirely a matter of chance. It may happen that for generations no animals carrying the recessive gene are mated with each other or that with two animals that carry the gene, one of them has the dominant gene present, which suppresses the recessive gene.
A breeder may mate only pairs of natural grey (agouti) rabbits for generations – the natural wild colour is dominant – but suddenly be astonished by an all black litter. Black is recessive by comparison with agouti. It may have taken a long time but

Syrian hamster

American crested guinea-pig

tails, nose, and ears. Breeding between a full coloured rabbit, both of whose parents were entirely coloured, and a part albino, produces only full coloured offspring. The offspring have inherited the dominant gene for coloured coats from one parent and the recessive gene that causes partial albino coat and eye.
These offspring therefore carry the albino gene, even though this cannot be seen because the gene that provides information for a coloured coat is dominant. These rabbits therefore carry a specific characteristic. If a brother and sister from this litter were to mate, there is a chance that some of the offspring would receive the recessive gene from both the father and mother, resulting in a number of partially albino rabbits.
The characteristic that could not be seen in the parents has come to light by the combination of two recessive genes. Mating between two rabbits that have coats that are full coloured can therefore produce partially albino rabbits but the reverse is not possible.

eventually, by chance, two animals have been mated that both carry the recessive gene, that have come together during the fertilisation of the egg. The chances of this happening are small but always possible.

True bred

Animals with which it is known for certain that they do not contain any of the potential surprises described above are regarded as true bred. In the language of genetics, such specimens are termed homozygote or are said to be homozygous for their colour and type of coat. An animal that exhibits the imprinting of a recessive gene is therefore always homozygous, whereas with animals which exhibit characteristics of a dominant gene, we can never be one hundred per cent certain that they are homozygous.

For example, black is a dominant gene by comparison with brown so, with a black animal where we know that one of the parents was brown, without doubt the animal must carry both the dominant gene for black of one parent and the recessive gene for brown from the other. Such animals cannot be true bred for black but in the language of genetics heterozygous.

Breeders and others simply say they carry the recessive gene, in this case for brown. Usually the intentional test crossing of such a carrier with a brown animal will produce brown offspring but because of the haphazard way that genes are brought together, a series of test crossings may be required. Despite this, there is a degree of order in terms of the inherited characteristics. This order really applies across the spread of large numbers but even in the smaller scale of one breeder's stock, there is a fair degree of order.

Genetic code and calculating the odds

Dominant and recessive genes are indicated by code letters. A dominant characteristic is always indicated by a capital letter. If we

Honey-coloured Mongolian gerbil.

take black as an example, this is a capital **B**. This colour dominates over a brown coat colour, which is simply designated by the small letter b. When a true bred black rabbit (**BB**) is mated with a true bred brown rabbit (**bb**), the offspring receive one gene from each parent. Their genetic code becomes **Bb**. The dominant presence of the gene for black coats will manifest itself in all animals. They do however carry, invisibly, the genetic information for a brown coat. If these offspring are in turn crossed, the following combinations are possible:

BB true bred for black;
Bb not true bred for black;
bb true bread for brown.

This means that there can be two dominant, or two recessive genes, or one dominant and one recessive gene.

The litter probably consists of a mixture of black and brown animals. Because the black gene is dominant over the brown gene, the chances are greatest that most of the litter will be born black. In terms of percentages, it is likely that 25% of the offspring will be true bred for black (**BB**), a further 25% are likely to be true bred for brown (**bb**) with the chances of non true bred black animals (**Bb**) being the highest. On average, about half of the litter will have the same genetic code as the parents.

Small Chinchilla rabbit.

Patched Syrian hamster

Genetics in practical use

The colour possibilities, together with the different markings and patterns (e.g. tan, streaked, tortoiseshell) are extremely wide. This is equally true of other characteristics, such as length and shape of tail, texture and length of the hairs in the coat, and the density of any undercoat. Not every characteris-

Dwarf Campbelli hamster

White mouse

tic is inherited so neatly, dominant or recessive, according to a book. There are characteristics that only reveal themselves when another gene is present that activates the working of a different gene. There are genes that only pass on their inherited properties in combination with other genes. A particular colour of eye, for instance, may be linked to a specific form of marking or pattern of coat, making it genetically impossible to breed that eye colour with a different type of coat. With both rabbits and rodents, there are genes that carry certain colours that are linked to particular illnesses or defects and some combinations which can never be true bred, because the offspring with two identical genes cannot survive or have a reduced chance of survival. A "lethal factor" comes into play that occurs, for example, with white mice. There is much to discover and to be discovered in the entire field.

For those who are interested, there are various publications that go into the subject of genetics with rabbits and rodents in much greater depth.

Gerbils

Tame rat with Siamese markings

Shows

The aim of showing animals

There are three aims of shows: firstly both breeders and enthusiasts are curious about the opinion of the judges, secondly breeders can compare their animals (it is the chance to see the animals of other breeders without having to spend time and money driving long distances to see them), and thirdly, a show has an important role in creating interest in rabbits and or rodents. Nowhere else provides the same assortment and number of animals to admire as at a show.

Shows may be small scale affairs, such as a local or regional show held in a village hall. In this case, the animals are judged by members of the association or club. There are then larger, national shows at which everyone who is a member of the appro-

Left: A Giant Chinchilla rabbit.

Dwarf Chinese hamsters in their original colour.

priate organisation or organisations may participate. Finally, there are big international shows. For anyone who is interested in the subject, an international show is the place to look at animals because of the sheer numbers of animals and different breeds. Almost every possible breed will be shown by breeders who may have travelled great distances.

If you wish to see a rare breed or unusual colour, and international show offers the best choice. Unfortunately for those in the United Kingdom, the travel restrictions on animals prevent international shows being held in those countries and British breeders cannot show their animals abroad.

What is a breed standard?

Animals are judged at shows against the standard for the breed or its specific variety. The standard is a written description for each breed or variety of how the ideal representative of the breed or variety should look. This carefully sets out such matters as the

length of the ears, the density of the hairs in the coat and its colour, and the weight of the animal.

The standard for a breed or its variety is published by the national governing body for the breed, in conjunction with the special breed clubs and associations.

At present there are no international breed standards so that a particular rabbit may have different weights in neighbouring countries. The colours and other conformational points may

also differ. For the standard for your rabbit, mouse, fancy rat, or hamster, the best place to check is the national breed organisation.

Guinea-pigs

Chinchilla

A judge assesses a mouse that has been entered.

The judge

A judge is a person who has been approved to judge one or more breeds at shows. Often this person will have reared the breed in question for many years or used to do so. Before passing an exam to become one, the potential judge has a fairly tough course of study to complete, covering genetics, diet and nutrition, and physiology.

It is therefore possible to expect that the judge not only has a detailed knowledge of the outward physical characteristics of the breed but also has substantial general knowledge about everything to do with breeding and the hobby.

Preparation for a show

The judging of a rodent or rabbit is a momentary consideration. The judge only see your animal for a few minutes at most and has to make an assessment in that short time as to its qualities.

If the animal is moulting at that moment, the judge will not take this into consideration but will undoubtedly take off points because the coat is not in optimal condition. This is why breeders and enthusiasts virtually only ever show animals that are in top condition. Claws that are too long also detract. Of course an animal needs to look its very best on the show bench. This is rather difficult

This Belgian Hare has learned to sit so that it shows to advantage.

with mice, hamsters, and fancy rats, because it is difficult to train them, so that show training really only applies to rabbits.

Imagine that a specimen of a breed that according to the breed standard should have a short back is stretched out on the bench; the judge is hardly able to form an opinion

When animals are moulting, as this Flemish Giant is, there is little point in entering them for shows.

about its back. Since a rabbit that is stretched out always appears longer than one sitting up straight, such a rabbit will get fewer points than its competitor that shows itself to best advantage.

Rabbits can be trained to hold the correct show posture that achieves the best show bench results. The more high-spirited breeds do this quite naturally and virtually no training is necessary. Other types of rabbit are more of a challenge.

Every serious breeder has a table in his or her establishment in which the rabbits can be regularly put in the correct position from a young age. It is best to have a piece of carpet on the table to give the rabbit something to grip on. Putting a rabbit into show posture requires some practice and experience.

This is one of the reasons why a beginner should initially only enter small local shows so that the art can be perfected from the experienced breeders.

2. Mice

Pet mice (*Mus musculus*)

History

The predecessors of our pet mice are the common house mice we might encounter in stables, sheds, and as the name implies, houses. The big difference between pet mice and the common house mouse is that pet mice have been bred and cared for by humans for a considerable time.

In that time, they have been further and further refined through selection so that their colours and bodies, quite apart from mutations, are different in a number of ways from their wild brothers and sisters. Keeping pet mice is not a hobby of recent origins. Writings show that pet mice were kept and bred in China and Japan long before the start of Christendom and white mice were kept captive on the Greek island of Crete in the years before Christ.

These mice were picked up in the hand and it is recorded that there were special temples with thousands of white mice in them that were maintained at state expense.

Left: Mice with satin coats.

Mice enjoy the company of other mice.

Three young mice.

The ancient Egyptians also kept mice in captivity and deemed them to have supernatural powers. Tame mice were used to foresee into the future. In more recent times, enthusiasts in Britain started selective breeding of mice during the nineteenth century to achieve specific colours and markings. The first breeding specimens to reach Britain did not come, as might be imagined, from Greece but were probably brought back from China by Portuguese seafarers.

The National Mouse Club (NMC) was founded in England at the end of the nineteenth century as the first organisation to concern itself with the breeding of pet mice. Today fancy mice are also bred throughout the world in laboratories but fortunately some also find a home as pets, because they are delightful, interesting, and easy-to-care-for rodents.

Behaviour

Mice are active, enterprising, and above all inquisitive creatures. During the day they are somewhat calmer and sleep a great deal, whereas they become much more active in the evening. They can be active during the day but with periods of rest in between. Mice are extremely expert climbers and make great use of their tail when climbing, winding it around things to hang on.

Mice can also jump extremely well. Apart from the occasional few, mice rarely bite their handlers.

Mice are gregarious creatures, just like their relatives living free. For this reason it is best to have a pair or more mice but if you do not want to breed them, it is best to have two or three females. Male mice are not so popular as pets because of the strong smell that they give off. Animals in a group are very tolerant of each other and fighting within the same group of mice is virtually unknown. This requires the mice to all have known each other from a young age.

If a mature mouse is added to an existing group, it will be seen as an interloper and chased away. The fighting that can result can be quite severe so it is best to prevent the situation. If you have a group of mice that get on happily with each other then do not remove one or two of them from the group for a day or two.

Mice do not recognise each other by colour or other external characteristic but principally by smell. An animal that is away from the others for a time will not be recognised by the group when it returns.

Young mice, even when they do not know each other, can be put together without any problems so that you should not expect difficulties if you put animals of about four weeks old together that come from different litters. Offspring that are born to mice within the group will always be accepted by the other mice in the group. This does not mean that it is impossible to put different adult mice together but it does need some understanding. If you put them all at the same time

Mice are extremely adept climbers.

Laboratory cages are often used but are not very spacious.

into a thoroughly cleaned cage and ensure there is plenty to divert them, such as plenty of things to nibble and new things to play with, they will be so busy that they have no time for attacking each other. Caution is however necessary and it best to stay close by at first so that you can deal with any problem if it arises.

There is usually a very clearly defined hierarchy between mice in a group, with the dominant females and males being the only ones that will reproduce if there is any stress for the group – such as too many mice in a small area. With the less dominant animals in this situation – which

should not occur with mice that are properly cared for – the natural hormone action is regulated to prevent them from breeding.

Accommodation

Mice are good climbers and jumpers and they can also get through the smallest of gaps or holes. These Houdinis among rodents therefore need to be kept in a cage specially designed for them. Good mice cages have bars that are not more than 6mm (½in) apart. Any other type of cage is unsuitable for keeping mice in because the bars are usually further apart.

Do not overlook the fact that mice are quite cunning. A door that can be opened by us with one finger will not be an unsurpassable obstacle to an enterprising mouse. The door needs some additional security such as a small catch to keep it in place. Laboratory cages are often used to house mice with their

Mice like to play and to investigate anything you place in their cage.

hard plastic bases and lid of very strong wire mesh securely fixed to the base. The disadvantage of this type of cage is that they are rather low and do not offer much freedom of movement.

In addition to specific mouse cages, old glass aquariums are also suitable for use with mice. Their advantages are that they are cheap – old looking ones can often be found at flea markets and boot sales for little money – and no mess falls out of the cage. There are two disadvantages with them though: they are heavy, making them more difficult to clean, and they do not provide much ventilation.

This means that urine gases remain trapped longer, especially when it is warm, the cage is not cleaned early enough, or there are too many animals in one cage.

The temperature of such a glass cage can also shoot up quickly if the sun shines directly on it. Provided these matters are taken into consideration, there is no reason for not using an aquarium as a cage for mice. Because mice are such good jumpers, it needs to have a heavy lid mesh lid that fits well.

Mice have another capability you must not underestimate. They can bite through any wooden surface and also those of softer plastics.

Cages that are partly made of soft plastic or wood are therefore best avoided.

Two mice

There are various materials that can be used to cover the bottom of the cage.

Many people use wood shavings but shredded paper or maize are also ideally suitable, as is biological cat litter. Mice like to have somewhere to hide away in their accommodation. This could be an upside down flowerpot with the hole enlarged, a cardboard box, a mouse or hamster house, or a bird nesting box, which can be bought from pet shops.

A mouse is an enterprising and inquisitive creature that would be bored to while away its days in a dull cage with just a layer of wood shavings. If you put something new into their cage each week, the mice will not only be more active, they will be happier. You might fix a length of sisal rope to the bars so that the mice can be acrobatic on it.

Bits of tree root, willow branches, or empty kitchen rolls are all suitable playthings. Be careful with treadmills because mice can get

Mice with Siamese markings and satin coat.

their tails caught in them, especially if there are other mice in the cage. This sort of toy needs to be used under supervision.

Food

Mice will eat virtually anything – they are omnivores. They will eat both vegetable and animal based feeds. A commercial rodent feed contains everything needed, although the mice will appreciate this being supplemented from time to time with fresh vegetables, fruit, and some animal protein or an insect.

Do not give mice too much green vegetable matter and limit those with fat, such as peanuts and sunflower seeds to modest quantities. Mice do not eat a great deal: on average about 10g (^{1}/$_3$oz) for each animal per day. Hay should always be available for them in sufficient quantity because fibre is very important for their digestion.

To fulfil their gnawing urges as rodents, place a willow branch or chewing block in their cage. A mineral block that you can find in pet shops gives a mice something to chew if they need to boost certain minerals. Make sure there is adequate fresh water, preferably from a drinking bottle.

Champagne and tan mouse with a white bottom.

Blue and tan mouse with white bottom.

Care

The frequency with which you need to clean out the cages is dependent on the amount of space the animals have, ventilation, number of animals, the type of cage, and the bedding material. On average it needs to be done about once each week. Once each moth on average, the cage will need to be disinfected. Mice keep themselves very clean and regularly groom themselves, They do not need any bodily care.

Handling

Mice become tame fairly quickly if they are used to being picked up from a young age. You can pick a mouse up by holding the base of the tail with one hand and supporting the body with the other. Never grab hold of a mouse by the middle or tip of its tail because this can damage the tail.

Mother mouse with offspring of different ages.

A mouse litter of about eight days old.

Sexing mice

The difference between the two genders can be clearly seen by the distance between the anus and sexual orifice. With females, the distance is much shorter than with males. Adult males emit a smell that most people find unpleasant. This is why virtually only female mice are kept as pets.

Reproduction

Both the house mouse and its close relative the pet mouse are prolific reproducers. Female mice are already sexually mature before they are three months old and if nature is permitted to take its course, a mice can produce about nine litters each year. This has its cost of course in terms of the health of the mother, the numbers in each litter, and a reduced vitality on the part of the offspring.
Do not mate a female mouse until she is about four months old. Female mice will not always allow themselves to be covered. They come into season briefly about once every four to six days – the cycle varies from mouse to mouse – and only then will they permit mating to take place.

Normally you can place the female with a male during the time she is in season without any skirmishing between them. The female starts to make a nest towards the end of the eighteen to twenty-one days of her pregnancy. The young are born blind and completely naked.

Silver-agouti satin-coated mouse.

Two golden-agouti mice – one with normal coat, the other is satin-coated.

Their numbers can vary from four to eleven and sometimes even more. The size of the litter is dependant on both the age of the female mouse and her condition. About three days after they are born, the young start to grow hair and a rough idea can be gained of their colour and any markings. The coat takes about ten days to develop fully. Four days after this, the offspring can start to eat some rodent feed.

Once the mice are three to four weeks old, they can be separated from their mother. It is advisable, in any event, to remove the males from the nest because they are sexually mature at five weeks old. The mother and her offspring should be given as much undisturbed rest as possible. Because mice are such sociable animals, females in season can be left with the rest of the group. Her young will be cared for by the other suckling mothers in the group.

Life expectancy

Unfortunately, mice do not live very long. The average life expectancy for a pet mouse is between one and two years.

Colours of pet mice

Agouti

Agouti mice have a ticked coat in which each hair has both light- and dark-coloured bands. With agouti-coloured mice, it is important that the ticking is as regularly distributed as possible across the entire body. Lighter or darker patches or other irregularities are undesirable. The belly too should have a regular ticking in its coat, although here it is less prominent then the rest of the body.

GOLDEN-AGOUTI
The original agouti colour is golden-agouti. This is the colour of wild house mice, albeit usually somewhat duller than with pet mice. With this colour, the darkest hair colour is black and the lighter colour is a warm golden brown. The undercolour of the hairs next to the body, is a dark slate blue. The eyes and ears are black, together with the tail and soles of the feet. The belly and legs are less strongly ticked.

SILVER-AGOUTI
Mice which have black ticking bands on their hairs, interchanged with silver-white ones, are known as silver-agouti. The base-

Silver-agouti mouse

Exceptionally fine looking white mouse.

coat colour is dark slate blue and, in common with golden-agouti mice, the ticking on legs and belly is less pronounced. The eyes are black, as are the whiskers. The claws, ears and tail are darker coloured.

Silver-agouti mice can also have a different colour than black in their ticking.

CINNAMON-AGOUTI
Cinnamon-agouti mice have cinnamon ticking with lighter-coloured colour banding than golden-agouti mice and the base-coat layer is also lighter.

Their eyes are dark coloured and the ticking on the legs and belly is less intensive than elsewhere on their body.

CHINCHILLA-AGOUTI
There is also chinchilla-agouti, which looks quite similar to silver agouti but has absolutely no ticking on the belly. The belly and insides of the legs are plain white, with a slate blue base-coat colour and black eyes.

ARGENTA OR YELLOW AGOUTI
The argente- or yellow-agouti is also a wild colour. The black pigment of the ticking is softened to a light grey, making it so light and barely noticeable that the animal

appears plain pale yellow. The white belly betrays the wild colour influence. Argentes have red eyes and little or no pigmentation on the ears, soles of the feet, and claws.

Self-coloured mice

Self-coloured or pure-coloured mice have a single colour. The belly may be somewhat lighter, or perhaps duller, together with the legs, but breeders and judges are striving after single-coloured mice that have the same colour on their entire body, without any subtle changes and without any other colour present in the hairs of the coat.

Ideally, the coat is the same colour to the root of the hair, with the base-coat colour being the same as the rest of the coat. A number of different colours are recognised.

WHITE WITH RED EYES (ALBINO)
The albino white mouse with pink eyes is the most popular pet mouse of them all by a long way. This is also the colour which generally gives the best bodily conformation. Breeders of this colour do not need to exclude their

mice with first-class body shape because the colour is too dull, their markings too irregular, or the ticking is uneven. The colour of white mice with pink eyes is virtually always fine, giving the breeders room to select stock for their body shape, large ears, and smooth, shiny coats.

The breed standards for white mice with pink eyes are therefore much stricter than for other pet mice.
For instance, these mice may not measure more than 13cm (5^1/$_8$in). The sole of the feet and nose should be flesh coloured.

WHITE WITH DARK EYES

The white mouse with dark eyes is also more critically judged than most other pet mice. In terms of their body form and size of their ears, these mice and the white mice with pink eyes are the finest looking pet mice. The body must be plain white with pink soles to the feet and black eyes.

Breeding mice with black (dark) eyes is not

A group of chocolate-brown mice.

Blue-coloured mouse

an easy matter because inherited aberrations occur.

BLACK

Black mice have lustrous, dark black coats and their eyes, ears, tail, and feet are also black. The belly is often slightly lighter coloured.

CHOCOLATE

Chocolate brown mice have a plain brown coat.
The soles of the feet and eyes are dark brown, together with the tail and ears. The belly is often less lustrous.

LILAC

Lilac coloured mice have a pale pinkish-grey tint. These animals have blue eyes with a reddish glow.

CHAMPAGNE

The colour champagne slightly resembles lilac although it is softer and warmer with a tendency towards beige. These animals always have pink eyes.

BLUE

Blue mice should have as even a dark-blue a coat as possible.
The belly is often less lustrous but the eyes are always dark blue.

DOVE GREY

Dove grey mice have a delicate, pastel-like light grey coat with pink eyes. The colour is much lighter than that of blue mice and tends towards a metallic hue. Sometimes animals are born with undesirable blue or reddish tinges in the coat.

ORANGE

Orange mice have a warm orange tinge to their coat that is paler than that of red mice. One difficulty encountered with breeding orange mice is that they sometimes produce offspring with black tipped or ticked hairs. Their eye colour is pink.

YELLOW

Yellow mice have a warm yellow coat. The eyes are pink with darker claws and whiskers. There are also yellow mice that have black eyes but these are less common.

RED

Red mice have a warm, dark red coat with dark brown eyes. The tail and ears are somewhat darker pigmentation.

CREAM

Cream mice have a soft ivory-coloured coat with dark eyes. Breeders attempt to breed them with as even and as light a colour as possible.

Mice with true-breeding colour markings

The colour markings indicated under this heading are those that can be easily reproduced through breeding. This means that when two animals with the same markings are mated, their offspring have identical markings.

TAN

Mice with tan markings are extremely popular because of the contrast that the tan colouring makes with the main colour. With

Blue and tan and black and tan mice.

tan markings the entire underside of the mouse, together with the insides of the legs, is lightly tinged with tan. The rest of the body, including the head is darker coloured. The contrast between the two colours should be as sharply defined as possible, with clear boundaries between the two colours being preferred. There are various colour combinations with tan. The most usual combinations, and the original ones, are black back and head with rust brown belly and inside legs. There are also blue, dove-grey, chocolate, and champagne mice with tan markings. The eye colour is derived from the normal eye colour for the colour on the back of the mouse. Blue mice with tan markings, for example, have blue eyes, black and tan have dark eyes, and dove-grey and tan mice have pink eyes.

SILVER FOX

Silver fox mice have markings just like the tan markings, except that there is absolutely no red pigment. The bellies and inside of the legs of silver fox mice are white. At present, silver fox mice are mainly available in black, blue, and chocolate but there is no reason why they should not be bred in almost any colour found in mice.

SIAMESE MARKINGS

Siamese mice have similar colouring and markings as Siamese cats with the pale coloured body with darker extremities such as the legs, ears, nose, and tail, of brown or blue. Those animals that have blue markings have cream bodies while the blue marked specimens have a very delicate pale blue. The contrast between the colour on the extremities and the main body is clearly pronounced. Siamese mice always have pink eyes.

Siamese markings on a mouse.

HIMALAYAN MARKINGS

Mice with Himalayan markings closely resemble Siamese mice, however, their body colour is not cream but plain white. These mice are part albino and only ever have pink eyes. The extremities can be black, blue, or chocolate, although any colour is possible.

Markings

The difference between mice with markings and those with true-breeding colour markings is in the curious way the markings are passed on. If two mice with virtually identical and perfect markings are bred with

Silver fox mice in black and chocolate.

Himalayan marked mouse.

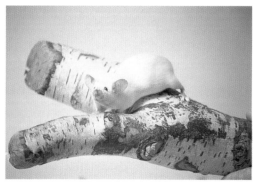

each other, it is possible their offspring will not bear the same markings. Breeding mice with good markings is quite a challenge for many and those mice that have good markings attract considerable interest at shows.

EVEN PATCHES

There are two types of patches for mice: irregular ones, known as "broken", and regular ones called "even." Mice with patches have a white body with a number of markings on it. Mice with even patches have patches which, when viewed from above, are symmetrical.

If the mouse has a patch on the left shoulder, there should be a corresponding one on the right shoulder. Patches can also appear on the head, particularly on the nose. The patches can be of various colours and two different coloured eyes are also possible.

BROKEN PATCHES

Broken patches are not symmetrical. The patches can be spread over the entire body but should not run together. There must be a patch on the nose. The patches can be a range of different colours.

TORTOISESHELL

Mice with tortoiseshell markings are two-coloured. The fields of colour are sharply defined with patches that are on average 4mm ($^3/_{16}$in) in diameter. The patches should be evenly distributed over the body. The ears and tail can be either single coloured or tortoiseshell but judges tend to prefer mice with tortoiseshell tails. One of the two colours is always cream, with the other being black or a colour derived from it, such as chocolate or blue.

Tortoiseshell marked mouse.

Mouse with Dutch markings.

Chocolate coloured mouse with white rump.

WHITE RUMP

This pattern has a colour at the front of the mouse's body, with the hindquarters being white. The dividing line between white and the other colour must be straight and the colours must not run into each other. White rump mice are best not bred with each other because the pattern is genetically linked to a lethal factor which causes some offspring to be still born or to die quickly. White rumps are bred in a variety of second colours.

Colour markings and patterns

Some breeders specialise in breeding mice that have both colour markings, such as tan markings, and patches, patterns, or banding. By cross-breeding white rump mice with tan marked mice it is possible to create animals with both these characteristics.

Special remarks

In addition to the colours, patterns, and markings described above, a number of new colours and markings appear accidentally from time to time.

DUTCH MARKINGS

Mice with Dutch markings have similar markings to Dutch rabbits: a coloured rear end and the same colour markings over both eyes. The rest of the body of these mice is white. The boundary between the white and the other colour should be sharply defined and as straight as possible. Various colour possibilities are possible, such as black, chocolate, but also various agouti colours are also recognised.

WHITE BANDED

It is not a simple matter to breed mice with a white band around their middles with the rest of the body coloured. The preference is for animals that have sharply defined and straight edged markings. The white banded mice can have a number of different colours as the other colour, such as black, blue, chocolate, and champagne.

White banded mice in blue and black.

Black and tan mice with white rumps

Sable coloured mouse.

Black and tan mice.

Black mouse with zebra markings.

Blue roan mice.

There are, for example, "zebra" mice which have two white bands around their bodies, and dappled or roan mice. It usually takes a few years before a new colour or distinctive marking is accepted by the various organizations.

Normal haired mice

Origins

The normal-haired mouse is the original pet mouse. These mice have the same type of hair in their coats as house mice. The wire-haired, satin-haired, and long-haired mice have all been bred from normal-haired mice.

Properties

The majority of mice shown have normal (short-haired) coats. Their bodily conformation is generally better than mice with the other hair varieties.

External characteristics

Pet mice have a slim and muscular build. The head should not be to pointed, while the ears should be well separated. large, and rounded. The line from the head into the body should flow smoothly. The preference tends to be given at shows to mice that are as large as possible but the overall impression should remain that of a mouse, and therefore not too robust.

An adult mouse that meets these requirements should measure about 13cm (5¹/₈in) excluding the tail. The tail should be as long as the body, thick at its base, tapering to a point. It must have no kinks in it or any other unevenness. Adult mice should weigh 40–60g (1½–2oz).
There is little different in size of weight between the sexes.

Coat

The normal-haired mouse coat lies smoothly against the body, is dense, and has a lustrous sheen.

Colours

Normal-haired mice are bred in every currently available colour, pattern, and marking.

Lilac mouse with broken patches.

The short, smooth hairs are ideal for coats with markings.

Satin-haired mice

Origins

Satin hair occurred spontaneously through mutation and was first written about in 1955.

Plain red mouse with satin hair.

Dove-grey and tan mouse with satin hair.

Siamese mouse with satin hair.

The mice with these coats made their appearance in a laboratory in Britain but the mice were soon discovered by enthusiasts. Mice with satin hair were first shown to the public at Carlisle, in the north of England, in 1973. Today, this coat variety is found in every country where pet mice are bred and kept.

Properties

Satin-haired mice – particularly those with normal-length coats – are increasingly popular. This is not to be wondered at considering the wonderful deep sheen that is associated with satin-haired coats, which show the mouse off at its best.

External characteristics

Satin-haired mice should have the same physical build as normal-haired mice.

Coat

The coats of mice with satin hair has a wonderful sheen, which heightens the intensity of its colour. With mice whose coat is out of condition – during a moult, or lacking in density, for instance – the sheen of the satin coat em-phasises the look of poor condition, making the coat appear to be greasy or even wet. This is particularly prevalent with the longer-haired varieties. Judges look critically at the intensity of the sheen at shows, which is a major feature of satin coats. Satin

Golden-agouti mouse with satin hair.

hair is bred with short-haired, wire-haired, and long-haired coats.

Colours

Satin-haired mice are available in every colour, pattern, and marking that occurs with normal-haired mice. The colour is usually more intense because of the satin coats sheen.

Wire-haired mice

Origins

Little is known about the origins of the wire-haired mouse. It is known that the first mice to shown this type of coat originated from a laboratory.

Chocolate-coloured wire-haired mice.

Long-haired white mouse with satin hair.

Properties

The coat of a wire-haired mouse does not require any special care or attention. The mice are also no different in character from normal-haired mice but unfortunately this attractive variety is still quite rare.

External characteristics

Wire-haired mice should have the same physical build as normal-haired mice.

Coat

The coat of wire-haired mice is slightly longer than that of normal-haired mice, with a definite sheen. The ideal wire-haired mouse has an erect crest or comb of hair that stands up from the centre of the back to the root of the tail, and at least a rosette of hair at each hip.

For shows, judges prefer to see more than these minimum number of rosettes.

Colours

Wire-haired mice are bred in all the usual colours and markings but the unusual coat looks best with plain and agouti colours.

Special remarks

Wire-haired mice are not yet as well established as normal- and satin-haired mice. It can therefore be difficult to acquire a wire-haired mouse. If you are interested in them, contact one of the organisations listed at the back of this book.

Long-haired mice

Origins

An Englishman, Tony Jones, was the first enthusiast to possess long-haired mice. He discovered them in a laboratory in 1966 and was able to take a couple to breed. Three years later, a breed standard was established for them in the United Kingdom.

Properties

A long-haired mouse does not require any special attention because the hair length is not comparable with that of the long-haired guinea-pigs. The character of these mice is the same as that of normal-haired mice.

External characteristics

Long-haired mice should have the same physical build as normal-haired mice.

Coat

The coat of long-haired mice should be as long as possible but retain its density. In reality, long-haired mice have coats that are slightly longer than normal-haired mice. True long-haired mice have not yet been bred but breeders are striving to create coats as long as they can.

Long-haired mice can also be bred with satin coats, and there are also long-haired

Long-haired white mouse with satin hair.

Long-haired dove-grey and tan mouse with satin hair.

Long-haired dove-grey and tan mouse with satin hair.

mice with curly hair, although these are very rare. The males tend to have longer coats than the female mice.

This is a logical consequence of the effect of the hormones on female mice, which also makes itself apparent in the quality of their coats.

Colours

Long-haired mice are bred in all the usual colours and markings. The longer hair is not ideal with the more complex colour markings and patches.

The colours cannot be as sharply defined with longer hair than with the shorter, sleek hair of a normal coat.

Special remarks

Long-haired mice are not yet well established and are not to be found at the local pet shop. It can therefore be difficult to acquire a wire-haired mouse.

If you are interested in them, contact one of the organisations listed at the back of this book or visit one of the bigger shows.

Hairless or naked mice

Origins

Hairless mice originated from a mutation that was established through selective breeding. Mice such as these are widely used for cancer research because the naked skin enables abnormalities to be seen more easily than with mice with normal coats.

Properties

Hairless mice are not very popular as pets. Not everyone appreciates the appearance of these unusual creatures. Apart from their remarkable appearance, hairless mice are identical to normal mice.

External characteristics

Hairless mice have the same physical build as mice with hair. They appear to have more wrinkles, however, and one of the disadvantages can be discovered if the animals injure themselves: scars and such things remain clearly visible.

Hairless mouse

Skin

The hairless mouse is totally without any hair.
They are born naked just like other mice and get hair after about two weeks but this gradually falls out when they are three weeks old and it does not grow back. Another characteristic of hairless mice is their thin skin. This is almost transparent so that the organs beneath are clearly visible.

Colours

So far the only colour of hairless mice is white, which manifests itself as a reddish pink.

Special remarks

Hairless mice have a greater need of warmth than mice with hair – which is rather to be expected.

These mice are bred in laboratories but also by enthusiasts but so far only in very small numbers. If you wish to obtain a hairless mouse, it is best to contact a mouse breeding or rodent association.

Adult hairless mouse and a young specimen, with which the hair has not completely vanished.

Two hairless mice.

African spiny mouse

Origins

The African spiny mouse is, like the house mouse, pet mouse, and rat, a true member of the mouse family or Murinae.

The most popular of these creatures for keeping in captivity and breeding are the Cairo spiny mouse (Acomis cahirinus) and the Black Nile spiny mouse which is a sub-species (Acomis cahirinus cahirinus).

Unlike the pet mouse and fancy rat, the spiny mouse is wholly undomesticated and these mice remain wild creatures. In addition to the two most popular types, there are also the Sinai spiny mouse and the golden spiny mouse. Most of these creatures originate from the northern parts of Africa but some are also found in the Middle East.

There is little rain and widespread desert throughout the area inhabited by these creatures so that their digestive system is adapted to a meagre diet and they can survive where there is no water because they derive sufficient moisture from grasses and insects.

The name "spiny mouse" can be simply explained by the fact that the upper part of their body is covered with spiny hairs. The remainder of their body has normal hair.

Behaviour

Although spiny mice are not domesticated, they still make first class pets. To keep them though requires somewhat more understanding of their needs than with the usual pet mice. Spiny mice are more adapted to sharing their lives with other spiny mice than with humans and show little interest in the person who takes care of them. Some spiny mice can become reasonably tame, while others remain extremely suspicious and ready to use their teeth if approached by a human hand.

Most creatures of the desert and semi-desert are active during the twilight and night, remaining out of sight by day, but many spiny mice are active during the day with periods of rest at intervals. Spiny mice are social animals that definitely need to be kept as a group.

Their life routines seem very similar to those of pet mice, which means that spiny mice that do not know each other are likely to fight if put together. If you want to keep a number of spiny mice, it is best to acquire a number of young ones and to then put them together in their new home. Any change to the status quo by introducing a newcomer will virtually always cause problems. Offspring from the group, by comparison, are always accepted.

Accommodation

Spiny mice can be kept indoors in a large glass cage or aquarium of which the top is covered with a tightly-fitting lid that also provides for ample ventilation.

It is best to choose a bedding material that is as natural as possible, such as clean sand, a few pebbles or stones, branches of non-poisonous wood, and some clean (dust-free) hay. A partially buried earthenware flower-pot can act as a burrow.

Food

Spiny mice happily eat rodent foods but need to have this supplemented by insects to feed on. You can catch and collect insects for them but there are also insect feeds for

Spiny mice have spiny hairs on their back.

A spiny mouse.

Spiny mouse.

Black Nile spiny mouse.

Sexing

The difference between the sexes is easy to see with spiny mice. The distance between the orifice of the sexual organs and the anus is greater with males than females.

Reproduction

Virtually all types of spiny mice have the same reproductive cycle. The average gestation period is 37 days.
The new born mice – in an interesting contrast with pet mice – are fully formed, with their eyes open, and they are quickly able to move about and fend for themselves.

After about a week, they start to nibble at food. Another difference from pet mice is that the litter consists of just two offspring. One way in which the spiny mouse is similar to pet mice is that they both raise their young within the group in complete harmony.

Life expectancy

The average life expectancy for a spiny mouse is about two to three years.

sale in the better pet shops. Do not feed them too much green vegetable matter because their digestive system is not accustomed to food that is high in moisture and this can lead to problems for them.

Care

The care of a spiny mouse is similar to that of a pet mouse. The cage must be cleaned regularly and new bedding must be provided at frequent intervals.

Handling

Spiny mice are not domesticated so do not expect them to allow you to pick them up to cuddle and caress them. These are animals to watch and admire but this is highly rewarding in itself, given their interesting behaviour. These animals are less suitable for young children.

Harvest mouse/ European dwarf mouse

(Micromys minutus)

Origins

Another much loved, and widely kept member of the mouse family (Murinae) is the harvest mouse.

Known in some countries as the European dwarf mouse, it lives up to that name and is one of the smallest mice in the world. The species lives in the wild throughout Europe and in parts of central Asia.

Behaviour

Harvest mice are amusing and interesting animals, whose behaviour interests a lot of people. These mice are active at night and during the day, with regular rests.
They like to climb. People who have observed them in nature have discovered that often they do not descend to the ground for day after day.
They are inquisitive and friendly towards humans and specimens held in captivity rarely bite their handlers. These are out and out social creatures that have great difficulty coping with loneliness. It is therefore best to acquire a group of young mice to start yourself off.

A newcomer in the group is usually fairly quickly accepted. These mice are widely distributed in nature but it is not recommended that you capture wild specimens that find it very difficult or impossible to adapt to a life in captivity.

Accommodation

A large glass cage or aquarium of which the top is covered with a tightly-fitting lid that also provides for ample ventilation is ideal for housing these animals.
Do not forget that these mice are extremely tiny and can squeeze through the smallest opening so that the ventilation openings must be very small but there must therefore be many of them. Put a thick layer of sand on the bottom. Because harvest mice like to climb and to be active, they must be given the opportunity to do so with bits of interestingly shaped branches from a fruit or willow tree, tufts of grass, stones, and a piece of tree root.

Food

Harvest mice eat lots of seeds but also need animal protein in the form of insects and their larvae. You can feed them with a mixture of rodent food, bird seed, and insect feed.

The insect feed can be obtained in dried form or fresh. Harvest mice do not drink a great deal, especially when their food has sufficient moisture content

Care

The frequency with which the cage needs to be cleaned out depends entirely on its size and the number of animals house in it. Generally, it is sufficient to clean out the cage and replace the bedding once every two weeks

Handling

Harvest mice are very small and move extremely quickly, making them very difficult to catch. If it is absolutely necessary to remove one of them from its cage, it is best to cradle it with one hand, using the other hand to form a cover to prevent it escaping.

Sexing

The difference between the sexes is easy to see with these mice. The distance between the orifice of the sexual organs and the anus is greater with males than females.

Reproduction

Harvest mice are sexually mature when they are one and a half to two months old. They are then able, under the most favourable circumstances, to bear four to seven litters each year.

The gestation period is about three weeks and the average litter consists of four to five offspring. Expectant harvest mice build ingenious nests, using plant material, that are raised off the ground on a stalk of grass, reed, or corn.

The young are born and raised in the nest. The offspring are blind at birth but they leave the nest at about two weeks old and

mix with the other adult mice in their surroundings.

Life expectancy

The life expectancy of a harvest mouse is about two years.

External characteristics

Harvest mice (European dwarf mice) weight about 7g (½oz) when fully grown with an average body length of 6cm (2³/₈in), making them the smallest European mouse. Their coat has normal hair that tends to be slightly darker on the back and sides with most specimens that on the belly.

Harvest mouse

3. Rats

Fancy rats

History

The forerunners of the tame or fancy rat were brown rats (Rattus norwegicus). Brown rats originate from Asia and arrived in Europe by stowing away on trading ships. The first brown rat was recognised and described in Europe in 1728. Prior to this, there were not known to be any brown rats in Europe. The rats subsequently spread to Britain and the USA. The only way this can have occurred was by ship.

Brown rats live by preference in a wet or damp environment and are not averse to water, resulting in them getting dubbed "sewer rats." The brown rat was better able to adapt to its environment than the native European black rat (Rattus rattus) and it quickly pushed the native species aside.

The eradication of both the brown and black rat was the only consideration at that time among the people of Europe. Since today's poisons were not available, people used all manner of ingenious methods of controlling rats. Some of the breeds of terrier were specifically bred as rat-catchers. The dogs were trained to kill the rats with one lethal bite. The dogs needed to be very agile and ferocious for this work. This led to farmers

Left: blue fancy rat.

Black patched fancy rat.

wanting to pit their dogs against others to see which was best. This was done by constructing pits in which a dog was placed with a quantity of rats to see how quickly the dog could dispatch the rats. Hundreds and sometimes thousands of rats were let loose in these pits, many of which had been trapped by professional rat-catchers, but increasingly there were rats bred specially for the purpose.

These contests drew large crowds and much money was wagered on the outcome. Two well-known rat-catchers of these days were Londoners Jimmy Shaw and Jack Black. These two men were known to have caught rats with different colour coats during the first half of the nineteenth century. These were kept apart from the other rats and bred. Those that were white or patched, and any other rat that was different from the usual wild rat were not sold to the killing pits but sometimes were sold to individuals as pets or to laboratories for research.

Rats were eagerly received in laboratories where they were and still are used in research into cancer.

Although the tame or fancy rat is directly descended from the brown rat, there is now little comparison between them. Fancy rats have been bred in captivity for so long that they have become domesticated, resulting in the friendly, gentle-natured creatures of today. Selective breeding undoubtedly played a part in this, since researchers do not want unmanageable or throwback research animals. It can be assumed that the strain of rats that was bred solely consisted of tame and well-natured animals. The rats found their way from the laboratories, where they were used for cancer research and behavioural studies, to enthusiasts, who kept them as pets.

In 1901, fancy rats were admitted to shows in England and clubs held classes where they were judged for their appearance. One of the greatest of rat fanciers, an English-woman called Mrs Douglas, died in 1921. She had played a leading role in establishing rats as valued pets with the public. Interest waned severely after her death and there were fewer and fewer people who showed any interest.

The decline in interest was of long duration and it was not until the 1970s before fancy rats started to make a come-back as popular pets and show animals.

This time, judging by the current strong interest in these sociable and interesting animals, their popularity is to stay.

Behaviour

Rats are highly intelligent, lively creatures that form a close bond with their handler. They make fine playmates for children but adults too can very much enjoy their company.

Rats that have been well socialised and handled with loving care rarely bite but they do need a great deal of attention to prevent them from becoming lonely and developing behavioural disorders. A fancy rat kept on its own needs a minimum of an hour each day of attention from its handler. Since not everybody can find this time day after day, it is far better to have two or more rats, that will keep each other company.

If two young rats are put together at the same time – whether male or female – there is rarely any problem about them getting on with each other. Rats are capable of living in huge colonies with young and old, male and females living together without fighting.

The females frequently raise their offspring together. Do take care though about placing two adult males that do not yet know each other suddenly together because this is not always successful.

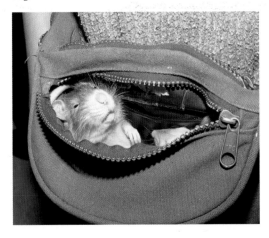

Fancy rats can be carried around with you in a bag.

Because of their intelligence, rats are quite able to learn, if sufficient time is devoted to teaching them.

Fancy rats are carried by enthusiasts on a shoulder or in a bag on their hip. It is a good idea to develop the bond in this way and rats can even be taught to recognise and respond to their own name.

Accommodation

Provided a rat can be regularly let out of its cage, then its cage does need to be so large. The size of its base should not be less than 400 x 600mm (about 16 x 24in) but it is far better to provide a rat with more space than this. Glass cages are as suitable as those of wire mesh although a stout mesh lid can

Husky rat

A spacious rat cage.

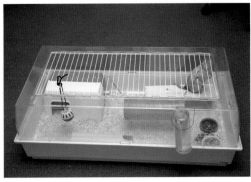

Rats are sociable animals.

A laboratory cage.

easily be pushed aside by these strong animals unless they are firmly fixed. A mesh lid is of course essential to ensure adequate ventilation.

The construction of the cage must prevent the rats from being able to slip through the bars or gnaw their way out. The cage must be easy to clean and to disinfect. The method of closing the cage needs careful consideration because rats are very intelligent and will soon work out how to open a door. An additional lock is not an unnecessary luxury.

The bottom of the cage should be covered with wood shavings but make sure these are not from pine because the aromatic nature of these shavings is not good for rat's health. Other less strongly smelling wood types are available, and also maize waste, or paper are better choices. Arrange a sleeping place in a corner of the cage, which can be made of wood, brick, or stone. Rats climb well and enjoy doing so.

Make sure there is something for them to scramble over and to climb on. Satisfy their inquisitive nature by regularly placing objects in their cage for them to play with and to investigate. Rats can easily become bored if they are not giving sufficient diversion. Checking out new objects in their domain provides a welcome change for them but make sure the item cannot be destroyed by chewing or that if it is, no harm can be done.

Rats eat anything; this one is eating an egg.

Food

Rats will eat anything, being omnivores who will leave virtually nothing untouched that is put before them. Make sure that you only feed them with a diet that is good for them, avoiding high calorie, fatty, and sugary cakes, sweets, and such like that are made for humans.

They can be given a mixed rodent food but they also need an occasional supplement of animal protein. This can be given in the form of dried cat or dog food or tinned pet food. Pieces of cooked chicken or beef are also suitable.

Other supplements to their diet include fruit and vegetables like apple, pear, banana, endive, carrot, and kale. Rats also like pieces of dried bread, yoghurt, and other healthy tit-bits. Always provide water in a drinking bottle because water trays can become soiled or get tipped over.

Two agouti rats and one with colour patches.

Tame rat with Siamese markings.

Care

Rats are clean animals that normally keep their coats clean themselves. It is not necessary to wash them very often but when you do, make sure that you use a neutral pH shampoo and then rub the animal thoroughly dry.

The claws can sometimes get rather long and require clipping with nail clippers. This is something you can do yourself carefully. If you are afraid of doing so, ask another rat enthusiast to do it for you, or get a vet to do it.

The cage needs to be cleaned out about once each week but this can sometimes be necessary more frequently – if

there are more animals – or it may not need doing this frequently. In any event, it is advisable to thoroughly clean the cage at least once a month and to disinfect it.

Handling

Fancy rats are very inquisitive creatures and when something interests them, they can throw caution to the wind. If you handle your rat well, then taming it will present no problems whatever.
To pick a rat up, catch hold of the tail by its root and support the rat with the other hand. Never pick it up by the tip or middle of its tail, since this can injure it.

Sexing

The difference between the two sexes is very easy to ascertain. The distance between the anus and the sexual orifice is much wider with males than females. Full-grown males are also larger and more calm than females.

Reproduction

Fancy rats are sexually mature at an age of about six to seven weeks but they are not sufficiently developed at that age to successfully rear a litter of offspring. Generally, a female can be mated at an age of about four months. A fancy rat comes into season about every four to five days so that a female rat can be mated once in each period of four to five days.

Females come into heat when there is a male in their surroundings and seldom come into heat if they are on their own. It is best to put the male and female together on neutral territory which does not bear the scent of either of them.

The male can remain with the female after the mating, during the pregnancy, and while the young are being reared but it is best to remove him to prevent the female from being covered again immediately after giving birth. The gestation period for rats is on average twenty-two days.

Female rats can happily coexist with each other, regardless of whether they are rearing young or not. They help each other to suckle the young and make no difference between their own offspring and those of another female rat.

ancy rats have large litters, with six to twelve being quite normal, although there are reports of rats giving birth to about twenty offspring! The birth weight of the babies is usually about 4g (¹/₇oz). The young are born totally naked and helpless, with their eyes closed.

The eyes open at about two weeks and the offspring start to get to know their surroundings. When they are four to five weeks old they have no more need of their mother and they can be separated from her.

A female rat becomes unable to produce young at about one year old and there is then little point in breeding her. The first litter is best delayed until the rat is about seven months old.

Life expectancy

The life expectancy of fancy rats depends on a number of factors. The quality of their feed plays an important role but their general health and condition is important too.

A life-span of two to three years is average and not exceptional, although rats can live to six years and older. The majority die young from cancer.

Fancy rats have an inborn susceptibility for cancer which is an inheritance from their breeding for laboratories.

Young husky rat.

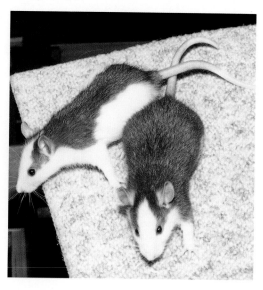

Colours of fancy rats

Agouti rats

The original colour for fancy rats is (golden) agouti.
The characteristic of this hair colour is the black banding or ticking on the hairs. In contrast with pet mice which are bred in a wide variety of agouti (or ticked) colours, fancy rats have so far only appeared with golden agouti, cinnamon agouti, and argente agouti coloured coats at European shows.

Masked rat.

Tame rat with original golden agouti colour.

A great range of colours is to be seen at shows in Britain and the USA.

GOLDEN AGOUTI

The alternative colour between the black bands or ticking is brown. The undercolour of the hair nearest the body is dark slate blue. The preference with agouti coats is for the ticking to be evenly distributed over the entire body, without any lighter or darker patches. The belly has less ticking than the back, head, and sides. The eyes are dark brown, and the ears, soles of the feet, and claws are dark coloured.

CINNAMON AGOUTI

Fancy rats with this colour have cinnamon-coloured ticking and the contrasting colour bands are a lighter brown. The belly is a pale grey-brown, while the whiskers are brown, and the eyes dark brown.

ARGENTE

Argente is a striking colour in which the usual black ticking is softened to a light grey. At first glance, argente rats look as though they are plain pale yellow with a creamy-coloured belly. The eyes are pink and there is little pigmentation in the soles of the feet, ears, and tail.

Self-coloured rats

Self-coloured rats are rats of a single pure colour. The belly and legs are generally a slightly lighter shade or less intensively coloured. Breeders strive, however, to breed

Self-coloured rat.

Tame white rat with pink eyes.

rats that are precisely the same colour over their entire body. Hair of a different colour and flecks do occur with self-coloured rats, although these are not really considered for showing. A broad range of plain colours is recognised for self-coloured rats.

BLACK

Tame self-colour black rats are plain black, preferably as dark as possible, without any tinge of brown, grey, or any other colour hairs. The eyes are dark brown, with the soles of the feet, claws, ears, and tail as dark as possible.

BROWN

Tame brown rats must be plain brown over their entire body. The tail, ears, soles of the feet, and claws are all brown, and the eyes are dark brown.

CHAMPAGNE

This colour has a tint of beige suffused with a rosy tinge. The belly and legs with champagne are usually lighter and the eyes are pink. The soles of the feet, nails, tail, and ears are coloured.

CREAM

Cream self-coloured rats have unpigmented soles to their feet, tail, and ears. Some have the associated pink eyes of these unpig-mented features but others have dark brown eyes.

ALBINO

Pure white rats with pink eyes and without any pigmentation in the soles of their feet, tail, and ears, are known as albino. The claws are a neutral horn colour.

Rat with Siamese markings.

Tame Irish rat.

WHITE WITH DARK EYES

In addition to the albino rat, there are also pure white rats that have slightly darker soles to the feet, tail, ears and claws. These rats have dark brown eyes.

Fancy rats with colour markings

There are certain colourations and markings that are relatively easy to breed true, so that when animals with the same colour markings are mated, there is assurance that the offspring will bear the same colours and markings.

SIAMESE

The colours of a Siamese rat are broadly similar to those of Siamese cats. The body is lighter in colour with darker extremities. The contrast between the body colour and the extremities is required to be as strong as possible. The eyes are always pink with this colour marking.

Siamese rats have not existed long but they have already become very popular.

HIMALAYAN
Rats that have part albino markings look a lot like Siamese rats but their body clear is pure white. This variety was first bred in France in the 1970s.

Markings

The position, size, and shape of the markings of fancy rats are different on every animal. Two specimens with perfect markings that are mated may produce offspring that comply with no standard but the reverse is also true.
Breeding rats with markings that please the judges is a great challenge for breeders

IRISH RAT
The Irish rat is a coloured rat with a large white triangular marking on its chest. The triangle should be clearly defined. The feet are white with unpigmented soles. Irish rats are bred in a wide range of colours.

BERKSHIRE RAT
Berkshire rats have white bellies. Ideally, they also have a white patch or blaze on their head too. The first Berkshire rats were shown at the end of the 1950s. The name is derived from a breed of pig that has patched markings.
Berkshire rats are bred in a variety of colours but the most usual variety is brown.

JAPANESE RAT
Japanese rats are without doubt the most popular rats with markings. It is extremely

Berkshire rat.

Tame rat with patches.

Silver-ticked yellow rat.

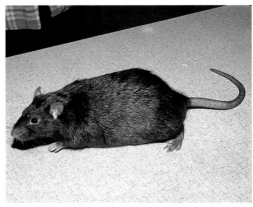

Silver-ticked black rat.

difficult to breed a Japanese rat with good markings.

The preferred form is to have coloured shoulders and head without any trace of white with a coloured stripe from the neck across the back to the start of the tail. The

An exceptionally fine Japanese rat.

dark pigment should run right through to the tail.
The remainder of the body is white. Japanese rats are bred in a wide range of colours.

SPOTTED

The preference at shows is for spotted rats to have markings that are as regular as possible.
This means that when viewed from above, the markings appear more or less symmetrical on either side of the body. A clear white head is highly regarded as are spots or patches on the belly. An eel stripe across the back should be as narrow as possible.

Rats with silver ticking

Rats that have a silver tinge to their coats, have silver-white tips to the hairs. The preference is for uniform silvering of the coat.

Fancy rats with silver ticking are bred in a number of colours, including brown and argente.

Special remarks

In addition to those colourings and markings already described, new colours and markings regularly appear on the scene. These arise from mutations, cross-breeding, and through continued selection to achieve a desired colour.

There are a number of these new arrivals that have not yet achieved overall approval at international level. One of the most recent of these "discoveries" is the Husky rat which immediately became very popular. Husky rats have a white belly and a white blaze. The upper part of their bodies is light grey. These rats are born dark-coloured but lighten over several months to their final colour. Another new arrival is the "masked" rat: a white rat with a coloured mask over the eyes.

A recent development has been the "capped" rat: white, with a coloured head. Self-coloured blue rats are gaining in popularity, as are lilac-coloured rats. There are more different colours recognised for rats in the USA than in Europe, although Britain also has a great variety. Colours such as coffee, walnut, amber, and lynx have been recognised in the United States.

Husky rat

Capped rat

Fancy rats with normal coats

Origins

The normal-haired coat is the original form. All other varieties of coat have been derived from the normal-haired variety.

Properties

Although rats have been bred with a range of different types of coat, the normal-haired varieties predominate.

External characteristics

The preference at shows is for powerful, muscular but slimly built rats with long bodies. Excluding the tail, an adult rat should be about 240mm (9½in) long and weigh about 500g (about 1lb).

The females are lighter and less robustly built than the males. The back and legs are slightly bowed. The tail is about 200mm (8in) long, thick at its root but tapers towards its tip. There should be no swellings or other blemishes on the tail.

The feet should be in proportion and lightly covered with hair. The front feet have four

Japanese rat

The smooth, sleek hair is ideal for showing off colour markings and patterns.

toes, while the rear ones have five. The ears and tail too are lightly covered with hair. A fine fancy rat has a relatively long head but the nose should not be too pointed.

The eyes are large and rounded and should give a lively impression. The ears should be wide, rounded, and stumpy, but not too large.

Coat

The hair of the normal-coated rat is short. The hairs should lay smoothly against the body with a sheen. The texture of a rat's coat is coarser than that of a mouse.

Colours

The normal-haired variety of rat is bred in the full range of colours and markings known to fancy rats.

Fancy rats with satin coats

Origins

The satin-haired variety of coat is a recent spontaneous mutation or sport which was discovered and described in the United States.

Properties

Fancy rats with satin-haired coats are not yet common place in Europe.

They are mainly found in the USA where they can be admired regularly at shows.

External characteristics

These rats have precisely the same body conformation as rats with normal-haired coats.

Coat

The iridescent sheen of the satin coat, together with its quality and density, are highly regarded at shows.

The coat consists of fine, short, and extremely dense hairs.

Colours

Satin-haired rats are bred in the same range of colours as normal-haired rats.

Fancy rats with rex coats

Origins

The first rex-coated fancy rat was bred by the British geneticist Roy Robinson in 1976. Rex coated rats were accepted by the English breed organisations that same year.

Properties

Rex rats only differ from other rats in terms of the texture of their hair.
The rex rat is no longer an unusual sight as they can be seen at most shows for small rodents.

External characteristic

Rex rats have precisely the same body conformation as rats with normal-haired coats.

Coat

Rex rats have curly-haired coats. The hairs of their coat are marginally longer than smooth-coated normal haired rats and the coat feels somewhat bristly to the touch. The whiskers of these animals are also curly. The adults tend to have more attractive curly coats than the young.

Colours

Rex rats are bred in all the usual colours of normal-coated rats. Because of the "untidy" nature of the curly coat, it is less suitable for showing off patterns and colour markings.

Sphinx (hairless) rats

Origins

The first hairless rat, which was known as a Sphinx rat, was reported in an American publication in 1932.

The variety resulted from a mutation to rats bred in a laboratory. The Sphinx rat is not

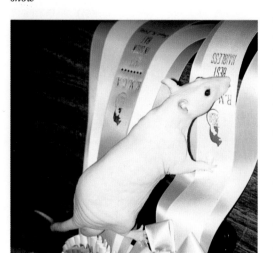

yet wholly recognised on the international scene and they are only shown in the USA.

Properties

The Sphinx rat has not become popular outside the United States so far. It is also unlikely that they will become widely popular because the naked appearance of these rats does not appeal to the majority of rat enthusiasts.

At first it was difficult to rear these rats to adulthood with most offspring dying because the female rats did not produce sufficient milk and had to be suckled by other rats to survive.

Through selective breeding, with only those hairless rats that produced milk normally, this negative characteristic has been bred out of the strain.

External characteristics

Hairless Sphinx rats have precisely the same body conformation as rats with normal-haired coats.
Some examples of the breed have strikingly wrinkled skin.

Skin

The hair of a Sphinx rat should be entirely without hair. A small amount of down on the head, belly, and legs is permitted. The skin needs to be very clean if the animal is to be shown, and free from any scars or blemishes. The skin is permitted to be wrinkled. Sphinx rats can have tiny, curly whiskers, or none at all: both types are acceptable at shows.

Colours

Sphinx rats are bred in the entire range of colours, yet because they are hairless, the colouring looks entirely different than with normally hairy rats.
These rats are judged by their markings rather than their colours.

Special remarks

There are two different types of hairless rat for the show judges: those that are entirely hairless and another variety that has tufts of hair which grow in constantly changing places. These are sometimes known as "patchwork" rats.

Fancy rats without tails (bobtail/Manx)

Origins

Bobtail or Manx rats arose from a mutation that was first discovered in the United States in 1942. The first reports were of tail-less rats born in laboratories. The first bobtail or Manx rat to be bred by a rat fancier was born in 1983 in the USA.

These rats were offspring from two Siamese rats that had been imported from Britain. Rats without tails are extremely rare outside USA and Britain and have not yet been recognised in international rat fanciers' circles. The fact that some litters contain rats that have difficulty with normal movement is of concern to many enthusiasts.

Properties

Tail-less rats are active, intelligent, and sociable. They climb and play like other rats and the majority of them are not in anyway troubled by their lack of a tail.

The exceptions are those specimens in which the lack of a tail also heralds other deformations in the bone structure, such as back and rear leg deformities. The more serious cases are extremely unpleasant for the rat.

External characteristics

Bobtail or Manx rats have a slightly different body shape than other rats with tails. Their bodies are somewhat pear-shaped, with the hindquarters noticeably higher than the front shoulders.

Only rats that have no tail at all are judged at shows, so show rats in this category must have no stump or part tail.

Coat

Bobtail or Manx rats are bred with normal, satin-haired, and with curly rex coats, and also without hair (Sphinx).

A rex-coated Manx rat.

The name Manx for tail-less rats is derived from the cat breed without a tail.

Colours

Tail-less rats are bred in the entire range of colours of fancy rats.

Special remarks

Some tail-less rats have difficulty with giving birth to their offspring. To prevent this problem, breeders normally only mate tail-less mails with rats with tails that carry the gene for no tail.

Pairing between two rats without tails or one without a tail and the other with can equally produce mixed litters of offspring with and without tails but also can give rise to some being born with a stump of a tail.

Dumbo rat

Origins

One of the newest developments with fancy rats is the Dumbo rat. This variety from California, USA, is named after the elephant in the Walt Disney film of the same name. Rats with a different ear form were first discovered there in 1991.

This mutation is so recent that it is not yet known how the gene will be passed on. Some publications talk of a singular reces-

Dumbo rat.

sive gene but since animals have been seen which have one normal ear and one variant, there can be no certainty.

Properties

It is not just the shape and size of the ear that differs from other rats with Dumbo rats, as they are at present known. The proportions of their bodies and characters are also different.

Dumbo rats are quieter and more gentle-natured than other rats. They have a shorter body and tend to be less alert and active. The coat of Dumbo rats can be normal, satin, rex, or hairless. Dumbo rats are bred in a wide range of colours.

External characteristics

The unusual shape of the ears is the typical breed characteristic. These are relatively thick and should be as large and as round as possible. The preference is for the ears to be low set.

The head is wider and more flattened than with other rats.
The body shape is somewhat similar to tailless rats, with a tendency to be pear-shaped and generally less slim than other rats.

Coat

Dumbo rats are bred in all the different varieties of coat but the original Dumbo rats were normal-haired (short).

Colours

The unusual shape of the ears and way they are attached to the head has no bearing on the colours.

The original Dumbo rats had a patch on their heads but every manner of colour has been created through cross-breeding with other strains of rat.

4. Gerbils (Gerbilae)

Mongolian gerbil *(Meriones unguiculatus)*

History

Mongolian gerbils are also often known as desert rats, although this name is quite wrong. Mongolian gerbils are actually more related to hamsters and voles than rats and mice *(Muridae)*. Mongolian gerbils are found in nature in the wide-open deserts of Mongolia and northern China. These are arid areas with very scarce rainfall and equally meagre vegetation. As a result of this, Mongolian gerbils are accustomed to a low protein diet and their digestive system is adapted to this.

In order to protect themselves both from predators and the high temperature differences which occur in desert areas, these gerbils spend much of their time in underground passages that they dig. The first Mongolian gerbils were exported to Paris from northern China in the middle of the nineteenth century by a Mr. A. David. This was followed by further expeditions which captured Mongolian gerbils in order to bring them back to Europe.

Left: Gerbil of the original colour.

Albino Mongolian gerbil

These animals were initially kept in zoos and laboratories but eventually came into the hands of enthusiasts. At first the only colour for gerbils was their natural wild colour but gradually, further colours have been developed through breeding. During the past twenty years, the Mongolian gerbil has become one of most popular small rodents.

Behaviour

Almost all gerbils are of a sociable nature which means that it is very difficult to keep one on its own. This is equally true of the Mongolian gerbil. These creatures always live in groups in the wild and when captive, are happier if living in the company of other gerbils. If they do not have company, gerbils suffer from behavioural problems and pine away.

This means that you must always acquire a minimum of two gerbils but it is far better to have more than this as company for each other. Adult specimens that do not know each other will virtually always react aggressively towards each other and the fighting between them can be a fight to the death. For this reason it is onlyadvisable to introduce young animals to each other in the same enclosure. Animals that are born within a group are always accepted by that group. Mongolian gerbils recognise each other by smell. If members of the group are separated for a few days, there is a strong chance the group will no longer recognise them and fighting will ensue. For this reason, never take a gerbil away from its group unless absolutely essential.

The gerbil makes a companionable and interesting pet that will only bite its handler if it can see no other means of escape. A gerbil that is properly handling and treated will usually not bite its handler. The Mongolian gerbil is, first and foremost, an inquiitive animal which is therefore fairly quickly tamed. Little escapes their attention: they frequently stand on their hind legs to survey what is happening about them.

Unlike other rodents, they are not restricted to being active at night and are quite active during the day as well. They doze off about every two hours for a rest. Mongolian gerbils are not really pets to be cuddled – they are far too lively for that – but they can get used to being picked up and will then less readily jump out of the hand.

Accommodation

Mongolian gerbils are often kept in ordinary cages with bars but although they do not really suffer in this type of accommodation, they are better kept in a glass cage. These creatures like to dig and in order to make this possible for them, it is best to put sand at the bottom of their glass cage.

To keep the sand aerated, mix wood shavings and hay with it. The gerbils will soon dig burrows as extensively as the thickness of sand and area of the base of the cage permit. Gerbils are happiest with this type of

Silver agouti gerbils

accommodation and will reward you with lots of fascinating activity. Do not forget that gerbils are outstanding jumpers and could easily jump out of their cage if you do not take steps to prevent them.

The well ventilated lid must be secure enough to prevent them escaping and resistant to being gnawed by their teeth. A gerbil's cage must contain a place for them to sleep. This can be a wooden box such as a bird's nesting box, an upturned flowerpot, or even a hamster sleeping box of hard materials. Make sure there are things for them to climb such as a bit of tree root or some large stones.

Never put a tread-wheel in a Mongolian gerbil's cage because their tails are far too easily broken with the slightest bit of resistance. The tail does not grow back after such an injury. Mongolian gerbils like to clean themselves in clean white sand (not builder's sand), kept in a separate container in their cage.

A flowerpot makes an excellent hiding place for a Mongolian gerbil.

Mongolian gerbils like to bathe in sand.

Food

Gerbils are accustomed to a frugal diet. Each animal needs only about 10–15g (¹/₃–½oz) of food per day. Their digestive system is adapted to very small amounts of food and water and a rich diet of sunflower seeds and peanuts is not healthy for them. Mixed rodent food is ideal, interspersed with small amounts of protein such as mealworm and other larvae, and also small pieces of fruit and vegetables such as apple and broccoli.

Make sure there is also ample hay in the cage because fibre is important for digestion with all rodents. Although these animals rarely if ever drink water in the wild, a small drinking bottle is not out of place so that they can drink as they require.

Since Mongolian gerbils teeth grow throughout their lives, you should put some willow branches or those from fruit tree in their cage for them to gnaw.

Litter of Mongolian gerbils.

Care

Mongolian gerbils excrete very little by way of urine or droppings and usually choose the one or at most two places to deposit these. If this corner of the cage is cleaned out once per week, then it is sufficient if the rest of the cage is cleaned out once per month. If you have quite a lot of gerbils in a small space then the cage will need cleaning more frequently.

Gerbils do not give off any smell that our noses can detect so if the cage begins to smell, it needs cleaning more often. Remove any uneaten remnants of food on a daily basis.

Handling

Mongolian gerbils quickly become tame with careful handling because they are such inquisitive and good natured animals. If you wish to pick a Mongolian gerbil up, do this by the root of the tail next to the body while supporting the body with your other hand.

Never pick a Mongolian gerbil up by the tip or middle of its tail, or even at the at the root of the tail if you do not support it, because they easily lose their tail and it does not grow back. This is a mechanism to help them escape from the clutches of predators.

Sexing

The difference between male and female Mongolian gerbils can easily be ascertained by comparing the differences between the anal and sexual orifices. The distance between them is wider with males than females.

Reproduction

Gerbils are sexually mature when they are about three to four months old and remain able to reproduce until about one and a half years old. The females are able to mate once every six days. The gestation period is about twenty-four days, after which the offspring are born blind and without any hair.

At birth, the young are about 20mm (_in) long and they weight 2–3g ($^1/_{14}$–$^1/_{100}$z). Gerbils give birth to their young in a sheltered place they have prepared in advance, such as a hollow they have excavated and covered with hay and fragments of paper. The average litter is between four to five young.

The offspring grow extremely quickly, developing their coat after a week and opening their eyes before they are two weeks old. At about this time, they leave the nest with their mother to get to know their surroundings. Once they are about three weeks old, they have virtually stopped suckling but it is best not to remove them from their mother for another week or two. Since Mongolian gerbils are extremely sociable animals, you can leave the mother and her litter with the other gerbils of their group. The male plays an important role in raising the young and should therefore not be removed during the rearing unless it becomes essential to do so.

Life expectancy

The life expectancy of Mongolian gerbils is between three and five years.

External appearance

Mongolian gerbils weigh 75–120g (2½–4½oz). The males are generally the heaviest. The length of their bodies, excluding the tail is about 120mm (4½in) with the tail itself measuring 80–110mm ($3^1/_8$–$4^3/_8$in).

Mongolian gerbils have a slender body with a very short neck. The body and relatively short tail are both covered with hair. The rear legs are longer than the front legs and the head is short and wide with a bent over back of the nose. The snout should not be pointed.

The relatively large, round, expressive eyes are the first thing that one notices about the gerbil. The ears are small and oval. Mongolian gerbils do not smell.

Coat

The hair of the coat is short, laying smoothly against the body, with a sheen.

Young Mongolian gerbil.

Colours of Mongolian gerbils

Agouti gerbils

Agouti gerbils have alternate dark and light bands on their hairs. This colour ticking is found in the natural or wild colourings of rodents.

At shows, judges look for sharp delineation in contrast between the coloured parts of the body and the white of the belly. There are various agouti colours, some of which are widely known but others which are more unusual.

AGOUTI OR NATURAL

The standard agouti is the natural original colour of the Mongolian gerbil. These gerbils have black and beige/brown banding on the hairs of their coat known as ticking. The base colour nearest the skin is a dark blue-

Mongolian gerbil with original colouring.

grey: the belly is always white. The eyes and the pads are black, as are the claws.

SILVER-AGOUTI OR CHINCHILLA
Silver agouti or chinchilla gerbils closely resemble the original colouring but instead of beige/brown colour banding, these specimens have white with the black ticking. These silver-agouti or chinchilla gerbils also have black eyes, claws, and soles of their feet. The belly is white.

ARGENTE
This yellow agouti colour is very popular. The belly of these specimens is always white and their eyes are pink, with the accompanying unpigmented claws and pads, and little pigment in the ears.

DARK-EYED HONEY OR ALGERIAN
The development of this colour is quite unusual. Dark-eyed honey-coloured Mongolian gerbils change colour at about two months old.

At first these animals have a plain pale yellow coat with black legs, tail, ears, and nose.
These markings of the extremities are somewhat like those of a Siamese cat. When they reach two months, the coat becomes warmer-toned and the black disappears from the extremities to be slowly replaced by black ticking in the hair.

The extent of the white on the belly increases with age. A dark-eyed honey-coloured gerbil also has dark claws and pads.

Argente Mongolian gerbil.

Honey coloured gerbil.

Self-coloured black Mongolian gerbil.

Lilac Mongolian gerbil.

Dove-grey Mongolian gerbil.

POLAR FOX

By cross-breeding dark-eyed honeys with chinchilla-coloured gerbils, breeders in Germany have created the polar fox colour. These gerbils are similar to honeys but everything that is yellowish with honeys is white with polar fox gerbils.

The change in appearance that occurs with honeys also happens with polar fox specimens. Their eyes are dark, as are the soles of the feet and claws.
Their bellies are white. The name for this colour is derived from its similarity with the colour of the polar fox.

Self-coloured Mongolian gerbils

Self-coloured Mongolian gerbils have the same colour over their entire body, without any ticking or white belly. Any white hairs or markings in the coat are undesirable but do occur. The colours of the eyes vary, depending on the colour of the coat.

BLACK

Pitch black Mongolian gerbils have a wonderful shine to their coat. The side of the belly is more matt. The claws, eyes, and pads are all also black.

LILAC

Lilac gerbils have a plain lilac coat, although in reality the colour is more of a pale blue-grey with a rosy glow. The side of the belly is less lustrous. The eyes are pink.

DOVE GREY

At first dove grey looks much like lilac but it is both lighter and cooler in tone. The eyes are pink.

WHITE WITH RED EYES

This is not an albino, as might first appear to be the case.
The body is pure white with pink eyes but unlike an albino, these specimens have a dark tinge on the tail.

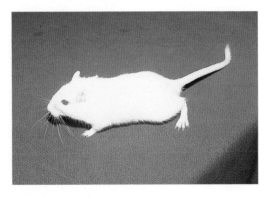

Other gerbil colours

SIAMESE
The markings of a Siamese gerbil are similar to those of a Siamese cats.
Just as with cats, the extremities of the legs, ears, nose, and tail are darker coloured, while the body is lighter.

BURMESE AND TONKINESE
The Burmese and Tonkinese colour mark-

Mongolian gerbil with Siamese markings.

ings also gain their name from their similarities with these breeds of cat. The Burmese and Tonkinese colour markings are similar to Siamese but are darker with a less marked contrast between the body colour and that of the points. The Burmese colouring is the darkest.

CANADIAN WHITE SPOT
Canadian white spots occur in a range of different body colours. These animals have a white patch or spot on the neck and head. Some specimens also have white legs and a white tip to their tails.

With otherwise plain colours (with no ticking or agouti colouring), there can also

Canadian white spot Mongolian gerbil.

Smoke Mongolian gerbil.

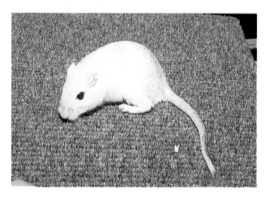

Smoke coloured Mongolian gerbil.

similar characteristics. Mating between two white spots can lead to smaller litters because some of the brood expire before birth.

PATCHED
Patched Mongolian gerbils have more white spots.

Special remarks

In addition to these colours, there are many new colours that have not yet been recognised internationally and there seems no end to the colour possibilities for the Mongolian gerbil.

Recent additions include white gerbils with dark tails, Himalayan gerbils, dark sepia, ivory-cream, cream, and anthracite colours.

Lilac coloured Mongolian gerbil.

Mongolian gerbils are inquisitive and attentive creatures.

sometimes be a white patch on the belly. The preference at shows is for white spots that are symmetrical. This means that breeding show perfect white spots is very difficult, since the spots vary widely in size and position.
The mating of two perfectly marked animals is no guarantee that the offspring will have

Other gerbils or jirds

The Mongolian gerbil is not the only gerbil kept as a pet but they are the most popular with the broadest range of coat colours. There are in the order of one hundred species of gerbil or jird.

Types that are encountered with gerbil fanciers include the pallid gerbil (*Gerbillus perpallidus*), the large and small Egyptian gerbils (*Gerbillus pyramidum* and *G. gerbillus*), Sundevall's jird, and Negev gerbil – of which several sub species exist, the fat-tailed gerbil (*Pachyuromys duprasi*) and Shaw's jird (Meriones shawi).

All these gerbils originate from the deserts and semi-desert areas of North Africa and the Middle East. The common link between all of these is that so far no colour mutations have been discovered. All of them are of a similar size with no more than 10–20mm ($^3/_8$–$^3/_4$in) between them.

The majority of gerbils have found their way to enthusiasts via universities and laboratories. The success of the Mongolian gerbil has led recently to increased interest in these other species. Since none of them has been kept as pets for long, there is far less information about them.

Furthermore there is also a debate underway between scientists about whether they are all correctly classified, which adds to the confusion caused by the fact that the similarities between species are so close that they are often confused with each other. Some appear so similar that only examination of their chromosomes can determine if they are the same species.

These other gerbils also make fine pets. It is expected that within a few decades, perhaps

Pallid gerbil

Negev gerbil

Negev gerbil

sooner, they will be as widely accepted as the Mongolian gerbil. For this reason, some of them are described below.

Duprasi's or fat-tailed gerbil (*Pachyuromys duprasi*)

Origins

The fat-tailed gerbil, which is also known as Duprasi's gerbil, originates from the northern parts of the Sahara desert in North Africa.

Properties

Fat-tailed gerbils are quiet, perhaps even lazy animals. They sleep almost the entire day and only go in search of food when night falls. They are less lively than other gerbils and limit their activity to the absolute necessities. They rarely bite their handlers. They only use their teeth if they are disturbed while sleeping or feel threatened.

Reproduction of these animals in captivity has not really proved successful. This is due the fact that these creatures are not sociable animals but live solitary lives. Some people have succeeded with keeping small groups of two to three in very spacious accommodation. The males and females need to meet each other in much the same way as Syrian hamsters.

The gestation period is about nineteen days. The young are born blind and without hair. The average size of litter is four. The offspring can fend for themselves within four weeks. The behaviour and reproductive habits of these animals is similar to Syrian hamsters but in the matter of their cage, it is better to follow the advice given for Mongolian gerbils.

External appearance

The fat-tailed gerbil is one of the cutest looking gerbils. Their rounded body is virtually without a neck, and the head is also broad. Their dark, oval eyes give them a friendly appearance.

The legs are quite short. Fat-tailed gerbils are 80–100mm (3½–4in) long and weigh 40–50g

Fat-tailed gerbil

Fat-tailed gerbil

The fat-tailed gerbil can store food in its tail.

Fat-tailed gerbil

(1½–1½oz). The hairs of their coat are short and soft, slightly longer than most gerbils, and somewhat greasy so that they stand-up rather than being smooth. The coat is sandy-coloured with a slight darker ticking and the belly is white. This colour is ideal camouflage for these creatures on the desert terrain where they live. The common name for these animals comes from the hairless stumpy fat tail which performs an important role for

Shaw's jird

these animals. They can store food and water in their tails in order to survive periods of scarcity.

Shaw's jird
(Meriones shawi)

Origins

The natural habitat for Shaw's jird is the desert areas of North Africa and parts of Egypt.

Properties

Shaw's jird, unlike most other gerbils and jirds, is not a sociable animal. They are quite intolerant of each other and it is therefore best not to keep them in mixed groups.

Female Shaw's jirds are especially renowned for their aggression towards males and they are not very friendly towards other females either. If you want to keep these jirds but do not wish to breed them, choose two males. Provided these have been used to each other from a young age, they usually get on with each other.

Both sexes of these animals are friendly towards their handler and rarely bite. The gestation period for these animals is about twenty-five days and the size of the litters vary from two to five.

Shaw's jird

Shaw's jird

Caring for Shaw's jird is otherwise identical to keeping Mongolian gerbils.

External characteristics

The body of Shaw's jird, measured from the tip of the nose to the start of the tail is about 120–140mm (4¾–5½in) long.
The short, lustrous hair lays smoothly against the body. Shaw's jirds are sandy coloured with a slight black ticking and white belly. The eyes, pads, and claws are darker. The tails is also covered with hair, with a black plume at its tip.

Pallid gerbils (*Gerbillus perpallidus*)

Origins

The pallid gerbil mainly originates from desert regions of Egypt but can also be found in other desert areas of North Africa.

Properties

Pallid gerbils closely resemble Mongolian gerbils and they can be cared for and housed in the same way.
These are active, inquisitive, and fascinating creatures, which can generally live happily in groups provided they are placed together when young. The pallid gerbil seems to have a greater need of warmth than the Mongolian gerbil. For this reason it is better to keep them indoors.

Pallid gerbil

Pallid gerbils have become increasingly popular as pets, in recent years.

Pallid gerbils

External characteristics

The pallid gerbil is often confused with the Egyptian gerbil. The two species closely resemble each other but the pallid gerbil is so far the more usual pet. These animals are smaller and thinner than Mongolian gerbils, measuring about 100mm (4in) from the tip of the nose to the root of the tail.

The head is fairly pointed and small so that the large, round, dark eyes and relatively big ears stand out even more. The hairs on the belly, legs, and underside of the tail are

Pallid gerbil

white while the rest of the body is a warm-toned sandy colour with a subtle darker ticking.

The colour provides excellent camouflage in their natural sandy habitat. The hairs are quite short and the hair on the tail is also shorter and less dense than the Mongolian gerbil.

5. Hamsters

Syrian hamster (*Mesocricetus auratus*) or golden hamster

History

There are quite a few different species of hamster but by far the most popular pet hamster is the Syrian or Golden hamster. These creatures originate from the desert areas of Syria.

The first reports of the species were made in the eighteenth century but the first tangible evidence of their existence was provided by the zoologist George Waterhouse. He made an expedition to Syria in 1839 and sent a hamster skeleton and skin to a London museum. It is thought that the British Consul, named Skene, brought the first live specimens back to Britain but no offspring can be traced from these animals. The Syrian hamsters that we know today are related, almost without exception, to the specimens that Professor Aharoni of the University of Jerusalem captured in 1930.

He sent offspring from these animals to laboratories, zoos, and universities in Britain and the USA. Subsequent visits to Syria by others resulted in further wild hamster breeding stock. The Syrian hamster eventually found its way from universities to individuals and soon became extremely popular. The world's first hamster club was formed in England in 1945.

The name of the Syrian hamster can cause some confusion. The original golden-coloured examples led to a popular common name of golden hamster and this name is still widespread. However, golden-coloured hamsters now form a minority of Syrian hamsters and serious enthusiasts generally prefer the name Syrian.

Behaviour

Syrian hamsters are solitary animals. They only depart from their solitary life style when it is essential for mating and rearing young but once the offspring are old enough to stand on their own two feet, there is no question of them all getting along harmoniously with each other. Young from the same litter that have spent weeks together suddenly start to act dreadfully towards each other. Syrian hamsters can be really fierce with each other if there is no means of escape. It does not matter whether the hamsters are male or female. Syrian hamsters must always therefore be kept in separate cages.

Syrian hamsters have scent glands on their flanks with which they mark their territory.

Left: the original colour of a Syrian hamster.

Two Syrian hamsters get acquainted.

Only young hamsters can stand each others company

The position of the glands can be recognised by the discoloration of the coat. Hamsters recognise each other through scent rather than by sight – they have poor eyesight – and also through the sounds that they make. Syrian hamsters are very much nocturnal creatures, sleeping for much of the day when they are seldom to be seen.

They roll themselves up in a sheltered hollow that they have made and sleep there virtually the entire day. Towards dusk, during the night too, and around dawn, they go in search of food. Everything edible is stuffed into the pouches in their cheeks and carried back to their hollow. The hollow is not just their sleeping place, it is also their larder.

If the ambient temperature remains below 10°C (50°F) for a while, and also when the days become shorter, they can hibernate. This hibernation is not an essential process for their health, indeed it can in some instances be dangerous for them. Syrian hamsters which are kept in centrally-heated homes virtually never hibernate.

Accommodation

Syrian hamsters can be kept either in cages with bars or glass containers. Suitable hamster cages have horizontal bars which allow the hamster to climb.

This type of cage also has ample ventilation but the disadvantage is that matter can fall through the bars to make the area around the cage messy. This problem does not exist with glass cages but ventilation is more dif-

The advantage of plastic and glass containers is that the surroundings remain clean.

Metal tread-wheel.

Dove-grey marked Syrian hamster.

ficult and the temperature can shoot up too high on hot days. Syrian hamsters sleep during the day and prefer to do so in a sheltered, dark place. Ready-made hamster cages are available that meet these requirements but a wooden nesting box for birds or a cardboard box can also be used.

Syrian hamsters can get by quite happily without such a sleeping place provided there is sufficient nesting material in their cage, such as hay, wood shavings, and ink free paper.

The hamster will make its own nest from these materials. The best bedding material for the bottom of a hamster's cage is wood shavings but not all types of shavings are suitable for hamsters.

Recent research has shown that small rodents can be allergic to wood shavings from pine trees. Be careful also with the use of newspaper because the ink can be poisonous.

Syrian hamsters make a hollow to sleep in.

Metal wheels are far better than plastic ones, since hamsters like to use their teeth. Make sure the wheel is large enough when you buy it and never place a wheel in a cage with a number of young hamsters because they can run on top of each other and injure themselves.

Some Syrian hamsters become so fond of the wheel that they become obsessive about it and are to be found on the wheel during almost every waking moment so that they exhaust themselves. With such hamsters, it is best to protect them from themselves by only putting the wheel in their cage for a couple of hours per day.

Hamsters are very meticulous animals that usually restrict their excreta to one or at most two places in their cage. You can place some cat litter in this corner so that the urine smell does not penetrate to the under tray or attack the sealant of a glass cage.

Syrian hamsters are inquisitive creatures that will investigate anything you place in their cage. Empty kitchen and toilet rolls will keep them busy for hours. A stump of wood, twigs from a fruit or willow tree, or interestingly shaped stones are all good to place in the cage.

Syrian hamsters need a reasonable amount of exercise so place a tread-wheel in their cage. The majority are happy to use this toy.

A piece of sisal rope that is held taught will be used to climb on by most hamsters.

Syrian hamsters eat both animal and vegetable food.

Food

There are various ready-mixed feeds for Syrian hamsters. A good hamster food contains few sunflower seeds, silage, and peanuts. Hamsters are not that partial to silage so that it is often left too long in their cage and although they adore sunflower seeds and peanuts, too much of these is not good for them because of the fat they contain.

In addition to a good quality rodent food, hamsters also enjoy fresh fruit and vegetables such as broccoli, carrot, and apple. Syrian hamsters can survive perfectly well on a low protein diet but they enjoy eating protein. Give them dried dog or cat food once or twice each week or some mealworms. Do not forget that hamsters need to gnaw at something. Provide them with a chewing block or meet this need with a willow branch. Make sure that some clean, dust-free hay and water are always available to them.

Care

Syrian hamsters are meticulous animals which keep their own coats clean. They do not need to be washed. Their claws can get too long and when this happens, they need to be clipped. The cage should be cleaned out about once per week and it is best to wash it thoroughly.

The "toilet" corner should be cleaned more frequently. Because Syrian hamsters store

Tri-coloured black Syrian hamster.

food away in their nest, you will need to check the cage regularly to make sure their is no rotting food. The water for a hamster should be changed daily even if they have not drunk any.

Handling

A Syrian hamster is picked up by cupping your hand (see illustration). This should be done carefully and never suddenly which could frighten the hamster and cause it to bite. Always approach your hamster at eye level and never from above. Leave a sleeping hamster alone to enjoy its rest.

Hamsters that are suckling their young will sometimes not want to be removed from their cage. In such cases, place a small box in the cage with some tasty tit-bits in it that the animal does not normally find in its feeding trough.

Horizontal bars are best for a hamster so that it can climb them.

The correct way to pick up a hamster.

Female hamster

Male hamster

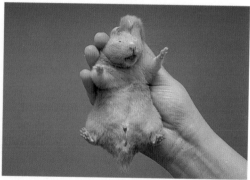

A day old litter of Syrian hamsters.

Almost every hamster will succumb to this temptation. While the hamster is filling its cheek pouches, you can move the box somewhere else while you attend to the cage. Syrian hamsters usually become very tame and are unlikely to bite their handler if they are properly handled.

Exceptions to this are older animals which have not been used to being held in the hand and being stroked. It is not always possible to gain the confidence of such specimens. The best way to let a Syrian hamster get used to you is to offer the animals food by hand as much as possible and approach it calmly.

Sexing

The difference between male and female hamsters can be determined by examining them beneath their tails.

The distance between the anus and sexual orifice for females is smaller than with males.

The area between is covered with hair with males but is bald with females. Adult hamsters only need to be viewed from above to see the difference between the genders. The females have rounded hindquarters whereas the males have more pointed rumps.

Reproduction

Syrian hamsters are often capable of reproducing from an age of about four to five weeks but the females are not ready at that age to bear and suckle young. It is far better to wait until the female is about four months old. The female is ready to be mated once every four days.

This can be checked by stroking her back: if she stands with her tail raised then she can be placed with a male that evening. Put both animals in a neutral cage that is unknown to them both and make sure the female is placed in the cage first.

The gestation period is about sixteen days with the young being born without hair, blind, and about 20mm (¾in) long. A hamster which is suckling her young needs as much rest as possible: too much stress can lead to cannibalism.

Limit your care and cleaning out of the cage to freshening up the toilet corner and providing fresh water. To inspect the litter, make sure that the female is first removed from the cage but do not use your hands to do this. If the female detects an unknown scent on her return, it can cause her to panic.

Hamsters which are rearing young can react in panic if they are disturbed too much and carry off their young.

Hamsters grow up quickly and can leave the nest at two weeks. By the time they are four weeks old, they no longer need their mother and can be put in a separate cages.

Life expectancy

The life expectancy of Syrian hamsters is two to three years.

Point of interest

The name hamster is widely adopted throughout the world. The origin of this name is derived from the German *hamstern* and Dutch *hamsteren* (to hoard), which refers to the way the hamster squirrels away food.

Syrian hamsters colours

Syrian hamsters with colour markings

GOLDEN

Golden is the original colour of Syrian hamsters and they are still widely known as Golden hamsters, although serious enthusiasts now term them Syrian. The golden Syrians have a golden brown coat with black ticking. The belly and insides of their legs are always ivory with dark eyes and ears. The base colour of the hair next to the skin is grey beneath both the coloured and ivory parts of the coat. A brown-black cheek flash marks the boundary between the golden-brown and the ivory.

GOLDEN WITH WHITE BELLY

This colour initially looks identical to the original hamster colour but the undercolour of the coat on the belly is white instead of the grey of the golden hamster. Specimens

Cinnamon hamster

Light grey hamster

Black-eyed cream Syrian hamster.

with this colouring are not normally mated together because such breeding has a high risk of deformities and early death.

YELLOW

Yellow hamsters have a warm yellow coat with black ticking and a black cheek flash. The undercolour is cream with an ivory belly. The ears are dark grey and the eyes are black.

Dark sable hamster.

SMOKE PEARL

Smoke pearl hamsters have a creamy grey coat ticked with black throughout and without an undercolour. The belly is ivory, there are black cheek flashes, black eyes, and grey ears.

LIGHT GREY/DARK GREY

Both types of grey hamster have a dark grey undercolour, ivory belly, black eyes, and grey ears. The light grey has a lighter coat and dark grey cheek flashes. The dark grey has a darker coat and black cheek flashes.

CINNAMON

Cinnamon hamsters have a ginger coat with a subtle brown ticking. The colour of the undercolour is light blue, with an ivory belly, brown cheek flashes, pink eyes, and flesh coloured ears.

LILAC

Lilac hamsters have a pale grey coat with soft lilac ticking. The belly is ivory and the undercolour is pale grey. The ears are flesh coloured, the cheek flashes are grey, and the eyes are pink.

RUST OR GUINEA GOLD

The rust or Guinea gold hamster has a pale brown ticking over a ginger coat.
The cheek flashes are dark rust brown and the eyes are also dark brown, with dark grey ears.

Self-coloured Syrian hamsters

ALBINO

Albino hamsters are entirely white with flesh-coloured ears. Their eyes are always pink.

WHITE WITH RED EYES

By contrast with albino hamsters, these specimens always have dark grey ears. The rest of their coat is sparkling white. They share pink eyes with albino hamsters.

WHITE WITH DARK EYES

These hamsters also look like albino specimens but these have black eyes instead of pink ones. The ears are flesh-coloured.

SABLE OR CHARCOAL

The sable has a dark-brown, almost black coat with a beige undercolour. The legs and rings around the eyes are also beige. The belly is the same colour as the rest of the body and the ears are dark, with black eyes. Some refer to this colour as charcoal. It was originally treated as a black.

SILVER GREY

Silver grey hamsters have a silver-grey coat with a grey undercolour.

Plain black Syrian hamster

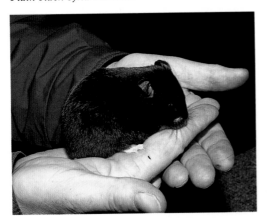

White banded cream Syrian hamster.

The cheek flashes are also grey and the ears are dark grey. The eyes are black.

RED EYED CREAM

The coat of these hamsters is a warm cream with a paler belly. Their ears are liver-coloured and the eyes are pink.

BLACK

Black hamsters should be pitch black and without hairs of any other colour in their coat but in reality most specimens have white feet and small white patches on their belly and throat.
The ears are often a lighter colour and the eyes are also black.

CHOCOLATE

Chocolate-coloured Syrian hamsters have a dark brown colour over their entire body. The ears are brown but the eyes are black.

CHAMPAGNE

Champagne is a soft grey tinged with lilac. Animals with this colour always have pink eyes.

COPPER

Copper Syrian hamsters have a copper coloured coat with which the belly is often lighter in tone. The ears are brownish brown. The eyes are ruby red.

DOVE GREY

Dove grey hamsters have a delicate shade of pale grey-blue. The eyes are pink.

Syrian hamsters with patterns

BANDED
Banded Syrian hamsters have a band of white around the middle of their back. The rest of the body is coloured. The preference is for bands that are clearly defined and with fairly straight edges and the ideal case is where the band's width is equal to about a third of the hamster's body length. Banded hamsters can have any manner of other body colour, including tortoiseshell. Banded offspring are only born when one or both parents are themselves banded.

SPOTTED
The first spotted hamsters were born in 1947. There is a range of different colours from coloured hamsters with the odd white spot, to dominant spot hamsters that are

Syrian hamster in a special cage for showing

Black-spotted dominant spot Syrian.

white with coloured markings. Judges preference is for hamsters with lots of spots, attractively distributed over the entire body. Spotted hamsters can have odd eyes (two different colours).

TORTOISESHELL
Tortoiseshell is a colour that is also found with cats. In common with cats, this colour is generally only found with females.
Tortoiseshell hamsters always have two different colours which are flecked throughout the body. The greater the number of spots and the more sharply defined they are, the higher is the standard of the marking. One of the two colours is always yellow hued, the other black or brown. These colours can be combined with white to give a tri-coloured coat.

Black tortoiseshell Syrian

119

Special remarks

The colours and patterns listed above are fairly common with Syrian hamsters but there are also many other colours and new ones are being constantly created through mutation, cross-breeding, and selection.

Short-haired Syrian hamsters

Origins

The short-haired coat is the original type for Syrian hamsters. The other coat types were all originally bred from short-haired stock.

External characteristics

Syrian hamsters have a wide, stubby head with relatively large eyes. The ears are erect. The fairly large cheek sacks, in which hamsters collect their food, are characteristic of the species. Hamsters are stockily built, being short and wide. The front feet each have four toes while the rear feet have five. Syrian hamsters are 120–160mm (4¾–6¼in) long, excluding their tails, which are 10–20mm (⅜–¾in) long. Syrian hamsters weigh 120–180g (4¼–6¼oz). The females are generally the largest and heaviest.

Short-haired golden and white Syrian hamster.

Dove grey tortoiseshell Syrian hamster.

Coat

Short-haired Syrian hamsters have a short and dense woolly coat, that is soft to the touch.

Colours

The short-haired Syrian hamsters are bred in the full range of colours.
The sleek coat of these hamsters is particularly suitable to show off the markings of tortoiseshells banded, and spotted hamsters.

Long-haired Syrian hamsters

Origins

The first long-haired Syrian hamster was born in the United States in 1972.
The genetic factor for long hair is recessive, meaning that two short-haired hamsters will never produce long-haired offspring.
If these offspring carry the recessive gene for long-hair though, then their young may contain one or more long-haired specimens.

External characteristics

With the exception of the long hair of their coats, the standard for these hamsters at shows is identical to that of short-haired types.

Coat

Long-haired Syrian hamsters have extremely soft hair that is ideally 20–30mm (¾–1¼in) long. The males generally have longer hair than the females. Specimens have been known to have hair up to 80mm (3⅛in) long but these are exceptional.

Most "long-hairs" only have longer tufts of hair in a few places – mainly around their bottom – while the rest of their body is short-haired. Caring for hamsters with longer-haired coats is more time consuming than for short-haired types. The hair tangles readily and can quickly look unkempt if it is not constantly groomed. Remove tangles by grooming with a toothbrush and very fine-toothed comb about once each week. It is best not to keep long-haired hamsters on sawdust because pieces can become trapped in the coat and cause tangles to form.

Long-haired Syrian hamster with a rare roan colouring.

Colours

Long-haired types of Syrian hamster are bred in the full range of colours and markings. The long hair lends itself best to self-coloured varieties because markings and spots will be less clearly defined with the longer hair.

A litter of Syrian hamsters can consist of many different colours and markings.

Many long-haired hamsters have longer hair predominantly over their bottom.

Long-haired Syrian hamster.

Satin-coated Syrian hamsters

Origins

The first reports of Syrian hamsters with satin-haired coats appeared in 1969. The type rapidly became popular.

The genetic factor for satin hair is dominant so that when a normal short-haired hamster is cross-bred with a satin-coated specimen, there is a strong possibility that satin-coated offspring will appear in the first generation.

External appearance

Satin-coated Syrian hamsters should have precisely the same body form as normal short-haired specimens.

Copper Syrian hamster with satin coat.

Coat

The satin-haired coat is short and should not be woolly.
The coat should be dense, with a wonderful lustre. Satin also occurs with both long-haired and rex coated hamsters.

Colours

Satin-coated Syrian hamsters occur in the full range of colours and markings. The colours appear more intense as a result of the satin lustre and red and black tints particularly appear to have a wonderfully deep sheen.

Golden Syrian hamster with satin coat.

Long-haired hamster with satin coat.

Long-haired rex Syrian hamster.

Short-haired rex Syrian hamster.

Special remarks

Breeding between two satin-haired hamsters usually leads to offspring with sparsely-haired coats. For this reason, satin coat types are always bred with normal short-hair hamsters.

Rex coat Syrian hamsters

Origins

It is uncertain how the rex coat came into being. What is known, is that the first rex types were shown in 1970.

External appearance

Rex coated Syrian hamsters should have precisely the same body form as normal short-haired specimens.

Coat

The hair is short, woolly, and wavy. The hair in the coat should be dense. The whiskers are also always wavy.

Colours

Rex coated types are bred in the entire range of Syrian hamster colours.

Special remarks

Cross-breeding between long-haired and rex types of Syrian hamster produces offspring with both characteristics. These animals have long coats that are slightly curly.

Rex Syrian hamster

Tri-coloured chocolate Syrian hamster

Dwarf Winter White Russian hamsters

(Phodopus sungorus sungorus)

History

The Dwarf Winter White Russian hamster is one of the most popular of the dwarf hamsters. Dwarf Winter White Russian hamsters were written about in 1770. They originate from northern Kazakhstan and Siberian. These animals have several different international names, including Siberian hamster and Winter White.

The latter name is derived from the fact that the animals have a virtually white coat in winter. The scientist Klaus Hofmann of the Max Planck Institute in Germany played a major role in the short history of these animals at laboratory animals and pets. Hofmann studied and bred these animals during the 1960s. The majority of Dwarf Winter White Russian hamsters with en-

thusiasts in Europe and North America originate from the animals bred at the Max Planck Institute.

Behaviour

The majority of hamsters are nocturnal creatures but the times that the Dwarf Winter White Russian hamster leaves its hollow vary widely: it might be at night-time, or in the daytime, or even a little of each. These dwarf hamsters are fairly calm animals, not skittish like the Chinese dwarf hamster or Roborovski hamster, but more thoughtful in their actions. This may explain their popularity.

These are not outstanding climbers. Smell is an important communication tool between themselves for Dwarf Winter White Russian hamsters and also plays a role in the relationship with their handler. They have scent glands on their belly with which they mark their territory. Its position can be recognised by the discoloration of the coat in its vicinity.
These hamsters recognise each other by their smell, not by sight or sound.

The Dwarf Winter White Russian hamster lives in small groups in the wild and they

Cream Syrian hamster

like to have company as household pets. This can be by keeping a pair or several of them in a small group. The best way to approximate their natural way of living is to have one male and several females. This of course leads to regular offspring and if you don not wish this, it is better to keep just two males.

Provided their are no females close by, these can usually live happily together. In common with other types of hamsters and gerbils, the group does not readily accept newcomers to its circle. For this reason, it is not advisable to attempt to introduce an adult to an existing group because it will be regarded as in interloper and chased away. It is not impossible to introduce adult specimens to each other though.

Place all the animals which are to form a new group in an unknown cage without any existing smell. These hamsters will usually not fight in such neutral territory which is not scent marked.

These Russian hamsters are clean animals, like all other hamsters. They clean themselves regularly to keep their coat in good condition. They normally choose one or two

Dwarf Winter White Russian hamsters are sociable.

corners of the cage which then become the only place or places they urinate. Their droppings, however, can be found throughout the cage.

Accommodation

Short-tailed dwarf hamsters do not jump and are not good climbers. They can therefore be kept in a glass cage without a lid.

A cage with bars needs more work to keep the area surrounding it clean and you will need to ensure that it is placed in a draught-free position.

The cage does not need to be very extensive for Dwarf Winter White Russian hamsters because they do not have a very active life style. You can place a tread-wheel in their cage to prevent them from becoming too fat. These creatures like to have a safe sleeping place available and this can either be a ready-made hamster house from a pet shop, or an upturned flowerpot or bird's nesting box.

Lay wood-shavings on the bottom of the cage but because of recent suggestions that pine shavings cause allergic reactions, make sure they are of a different type of wood. Make sure the hamsters always have ample hay because they use it for their sleeping quarters and like to eat it.

Straw is unsuitable as bedding because it is virtually non-absorbent.

Food

In the wild, Dwarf Winter White Russian hamsters eat mainly seeds and small insects. In their natural habitat, they rarely encounter fruit and vegetables so restrict the amount of these because they can cause diarrhoea.

Three different coloured Dwarf Russian hamsters.

Dwarf Winter White Russian hamsters can easily be tamed.

Carrots, apple, and pear are better for them than cucumber, lettuce, and green vegetables with high moisture content. Your Dwarf Winter White Russian hamster will get an adequate diet if you feed it a good quality mixed rodent food. The hamster can be treated occasionally with some animal protein in the form of small larvae and dried cat food. These Russian hamsters do not drink much but they must always have fresh water available to them.

A bowl of water will quickly be knocked over or covered with sawdust so a drinking bottle is a better solution.

Care

Clean the cage out completely once each week. If the animals have plenty of space available, then this can be left until every two weeks. The "toilet corner" will need cleaning more frequently – about every two to three days. The care of Dwarf Winter White Russian hamsters includes the daily replacement of water, regardless of whether the hamsters have drunk any and removal of any food remnants that have not been eaten so that they cannot rot.

Handling

Dwarf Winter White Russian hamsters can be tamed relatively easily. It is important to always approach the little creature at eye level and not from above. That they bite from time to time is not really surprising because tiny rodents have many enemies in nature and use they teeth as a defensive system. Your hamster needs to learn that its handler does not mean it any harm.

The best way to do this is to offer food by hand and always to approach the hamster calmly. You pick them up by cupping your hands, using one to support it and the other as a lid. Take care that the hamster does not jump out of your hands. Most hamsters have little sense of height.
It is also possible to remove a hamster that has not been tame from its cage without any risk of being bitten. Put a small box or container with lid in the cage with some tasty titbits. Few hamsters can resist the temptation to check the food out.

Sexing

The two sexes can be ascertained by checking the distance between the anus and the sexual orifice.
The distance between them is greater with males than with females. Adult males can be easily discerned by the skin covering the testes. It is also easy to notice that adult

Male Dwarf Russians also help to take care of their young.

Dwarf Russian female with her offspring.

females, when viewed from above, are noticeably rounded at the bottom, while the males have a more pointed rump.

Reproduction

Female Dwarf Winter White Russian hamsters are capable of breeding from four to five weeks old but it is far better to wait until they are three months old. The females become fertile once every four days when they are prepared to mate with a male. A female that is ready to mate will stand still apart from the rest and then lift her tail. If you keep a pair of hamsters, then the male is likely to mate with the female.

If the male and female are kept in different cages, then let them get to know each other on neutral terrain – that is a cage that neither of them know, and that has no recognisable smells.
Dwarf Winter White Russian hamsters are generally not so eager to fight with each other. It is normally possible to leave the pair together, even when the offspring are born.

The gestation period of Dwarf Russians is about nineteen days and the average litter is four to six young. The female rears her young in a nest. During the period that she is suckling, the mother must be left alone with as much peace as possible. Nursing mothers that are subjected to undue stress can eat their own young in panic.
Limit the cleaning of the cage to dealing with the "toilet corner" and the replenishment of food and water. During the second week, the young start to explore outside the

Dwarf Winter White Russian hamsters should be rounded like a ball.

next and begin to eat some rodent food. From about the third week, they are fully weaned but it does no harm to leave them for another week with their mother before separating them.

Do not overlook the fact that if the male is left in the cage with the female, she will have a further litter soon after her offspring are weaned.

Life expectancy

Dwarf Winter White Russian hamsters live for between two to three years.

External characteristics

Dwarf Winter White Russian hamsters are supposed to be more of less the shape of a ball or sphere, according to the breed standards. The body should have a diameter of about 70mm (2¾in).

The head is broad with rounded eyes and small, round ears which have a light covering of hair on their insides. These hamsters have short legs and five toes on each foot, the soles of which have hair on them. The Dwarf Winter White Russian hamsters form one of the short-tailed group of hamsters, together with the Roborovski and Dwarf Campbells Russian hamsters. Their tails are sparsely covered with hair and are only about so that they are barely visible beneath the woolly coat of the rump.

Dwarf Winter White Russian hamsters are small. In the wild, where they certainly get less to eat than in captivity, and where they have to run about to find their food, they

weigh 20–25g (approx. ¾oz). Those kept as pets frequently tip the scales at twice this weight. In common with all other hamsters, the Dwarf Winter White Russians have cheek pouches, although these are less pronounced than those of Syrian hamsters.

Coat

These hamsters have a short, dense coat of fine hairs which appears woolly.

Special remarks

The Dwarf Winter White Russian hamster is also known as the Djungarian or Siberian hamster. Djungarian is a collective name for both the Dwarf Winter White Russians and

Dwarf Winter White Russian hamster

A cross between natural-coloured Winter White and Campbells Dwarf Russian hamsters.

Dwarf Campbells Russians. The two species can be inter-bred with a proportion of their offspring being able to reproduce. Such cross-breeding is generally ill advised because they do not serve any useful purpose.

Colours of Dwarf Winter White Russian hamsters

Natural wild colour

The natural wild colour is the basis colour for the breed and remains the most prevalent colour for these hamsters. Typically, these hamsters are grey-brown with a brown-black dorsal stripe.

The bellies are light grey to white and the ears and eyes are both dark. Three bowed or arched markings are found on the side of these hamsters which separate the colour on the back from the belly and colour on the rest of the body.

The colours on the back and the belly should not run into each other because judges like to see sharp definition of the arched dividing lines. This colour is genetically dominant.

Sapphire Dwarf Winter White Russian

Pearl Dwarf Winter White Russian

Dwarf Winter White Russian

Sapphire

The first sapphire Dwarf Winter White Russian was born in Britain in 1988. Sapphire closely resembles the natural wild colour of these hamsters, although the coat is blue-grey instead of grey-brown. The dorsal stripe and other markings are dark blue instead of black. The eyes and ears are dark like the standard colour. The sapphire gene is recessive when compared with the gene for wild colour.

Pearl

The first known example of this colour saw the light of day in Britain in 1989. This attractive colour is characterised by a white coat with an even grey-black ticking throughout. The ears are light grey while the eyes are black. The dorsal stripe and arched

Original winter coat colour of the Dwarf Winter White Russian hamster.

Pearl Dwarf Winter White Russian

lines are black but are not usually as clearly defined as with other colours. The gene for this is colour is dominant in comparison with both the natural wild colour and sapphire.

Special remarks

These creatures do not hibernate during prolonged periods of cold weather. As the daylight hours reduce, the coat starts to turn slowly white and can become patched with white or even totally white. In the wild, the white winter coat acts as camouflage against the winter's snow.

Male hamsters of this species are not fertile while they have their winter coat. When kept as pets, these hamsters rarely turn white unless their surroundings are darkened.

Roborovski hamster (Phodopus roborovskii)

Origins

The Roborovski hamster is a dwarf hamster from the farthest reaches of southern and western Mongolia where it lives in a desert habitat. There is little to eat and this hamster can, in common with other dwarf hamster, live on very meagre rations. Roborovski hamsters are a quite recent introduction as pets. The first examples made the step from laboratory animals to household pet during the 1970s.

Behaviour

Roborovski dwarf hamsters are short-tailed hamsters and like others of this group, they naturally live as a pair or in small groups. They are grateful of the company of one of their own kind too in captivity. You can consider keeping a male with one or more females.

These hamsters do not reproduce at the same rate as other hamsters, so you do not need to worry about overcrowding.

Roborovski hamster

Roborovski hamster

Roborovski hamsters are lively creatures that are most active in the evening, night-time, and early morning. In spite of their highly active ways, they do not climb much, being more adapted to life on the ground. Roborovski hamsters are not only sociable towards each other, they are friendly with humans and they are not inclined to bite.

This can of course depend on the way in which they are treated and approached. Because they are so lively, these hamsters are less suitable cuddly pets for children, although there is always plenty of activity to see in their cages. Roborovski hamsters are clean animals that always urinate in one corner of their cage, although their dropping can be found throughout the cage.

Accommodation

Roborovski hamsters can jump but not very high so that you can readily keep them in a glass cage without a lid. They can also be kept in a hamster cage but because of their size, a mouse cage, in which the bars are

Roborovski hamster

closer together, is more suitable. Young Roborovski hamsters in particular can squeeze between bars. The bottom of the cage can be strewn with wood shavings. Roborovski hamsters are active creatures that like to run around a great deal, so a cage that is some-what larger than the usual hamster cage is better for them.
A tread-wheel is worth buying but some hamsters are so taken with this activity that they can become obsessive, spending hours every day on the wheel. If you have such a

hamster, then remove the wheel from time to time to let the tiny creature rest. Make sure there is somewhere sheltered for the hamsters to sleep and rest. This can be a ready-made hamster house from a pet shop but a bird's nesting box will do just as well, or a hollowed out log such as the kind used for birds. Place some hay in the sleeping chamber.

Roborovski hamster

Food

Roborovski hamsters need very little food but this is true of all the dwarf hamsters. These animals have evolved to survive in areas where food is very scarce. Animals in captivity are more likely to become obese than to have too little food.

A well balanced rodent food that contains many different ingredients makes an excellent basic food for Roborovski hamsters. This can be supplemented occasionally with some dry brown bread. Fruit and vegetables – especially those with a high moisture content – should be given only sparingly, and

Roborovski hamsters are one of the most active types of hamster.

that includes cabbage and similar green vegetables.

These foods can cause diarrhoea. These hamsters need to have hay in their cage at all times. Roborovski hamsters need to have some animal protein from time to time and you can give them this in the form of mealworms, earthworms, and larvae about once each week. They also like dried cat or dog food.

Care

The frequency with which the cage needs to be cleaned out depends on the size of the cage and number of animals kept in it. This can vary from once a week for a smaller cage or a larger group to once every fortnight for a spacious cage with only a few hamsters. The "toilet corner" needs to be cleaned out about twice per week. Because these animals store food in their "nest", you will need to remove any uneaten food regularly to prevent any stale food from rotting.

Handling

These hamsters move rapidly and are not easily picked up. They are fun to watch but less suitable as cuddly pets. When you wish to catch one of these hamsters, place a box in the cage with some tasty tit-bits. Curiosity usually wins over caution so that you can move the hamster carefully to another place. Make sure you have some-thing to close off the box, or the hamster will jump out.

Sexing

The difference between the young male and female hamsters can be ascertained by checking the distance between the anus and sexual orifice. These are farther apart with males than females. Adult males can be recognised by their testes.

Reproduction

The Roborovski hamster is one of the most difficult of the dwarf hamsters to breed. There are several reasons. Firstly, these hamsters are quite choosy about their partners and if they do not like each other, they are unlikely to produce any offspring. These hamsters also produce fewer litters and most females do not give birth until after the winter. The young are very small and must certainly remain with their mother for five weeks, which is two weeks longer than with other dwarf hamsters. A further difficulty is that hamsters do not live longer than two years. Breeding is therefore not straightforward, but given the comparative rarity of them, it is certainly worth while for those who are not easily put off. In other respects, breeding Roborovski hamsters is similar to Dwarf Winter White Russian hamsters with the exception that the young must not be taken away from their mother until five weeks old. The gestation period is twenty-one days.

Life expectancy

The life expectancy of Roborovski hamsters is about one and a half to two years.

External characteristics

Roborovski hamsters are the smallest dwarf hamsters. These little animals measure 70–90mm (2¾–3½in). They have a short, broad head, with large round eyes. The tail, which is sparsely covered with hair, is hardly visible at 5–10mm ($^3/_{16}$– $^3/_8$in). The ears are proportionally quite large and stand erect. The soles of the feet are covered with hair. Adult Roborovski hamsters that are kept as pets weigh about 35g (1¼oz).

Coat

The coat of a Roborovski hamster is very soft and longer than most other dwarf hamsters. The hairs do not lay smoothly but stand up slightly to keep a ruffled appearance. At present there are no other varieties of coat for Roborovski hamsters but this is probably only a matter of time.

Colour

The natural colour of the breed in the wild is so far the only colour known for Roborovski hamsters. The coat is golden brown with a grey undercolour. There is not dorsal stripe or arched markings as with other dwarf hamsters. Lightly coloured markings above the eyes provide some interesting details. The belly is white. These hamsters are somewhat more grey in winter than summer.

Dwarf Campbells Russian hamster
(Phodopus sungoris campbelli)

Origins

Dwarf Campbells Russian hamsters originate from Central Asia, northern Russia, and northern China. The species was discovered

Argente dwarf Campbells Russian hamster

by Thomas Campbell in 1905 and is closely related to the other dwarf Russian hamsters. For a time it was even thought that they were the same species because of chromosome similarities and a capability to an extent to inter-breed.

The terrain of the two species does not overlap, however, and although there are similarities in chromosomes, the animals also have distinct differences in chromosomes. Campbells Russian and the Winter White Russian are both Djungarian hamsters. The Campbells Russian made their way to their lives as pets by the same route as most rodents. At first they were bred in laboratories for scientific purposes and then became known to enthusiasts. They were introduced into the pet market in the United Kingdom in the 1970s.

Behaviour

Dwarf Campbells Russian hamsters are mainly active during the twilight. They sleep

Natural colour but patched Campbells Russian.

Dwarf Campbells Russian hamster a few days old.

during the day. They are not really climbers, preferring to remain on the ground. Do not keep a dwarf Campbells Russian hamster on its own because they like to have company of their own kind.

They recognise each other by smell because their sight and hearing is less acute. They also have no sense of height, in common with other hamsters.

This hamster is the only one that to an extent pairs for life. If a male and female take to each other, they will show little interest in other hamsters of the opposite sex. It is best to keep them as a pair. These hamsters are quite clean, using just one or two corners of their cage to excrete in.

Accommodation

Campbells Russians can be kept in a hamster cage or glass cage. The advantage of a glass cage is that the surrounding area remains clean. Spread a good layer of wood shavings on the bottom of the cage and there must also be hay available in the Campbells Russian's home.

Try to avoid shavings from pine because they are not good for small rodents. Consider an exercise wheel for the hamsters to help keep it occupied and exercised. Make sure too that there is somewhere the hamster can withdraw to sleep during the day.

There are ready-made hamster houses in pet shops of wood or solid materials, but a wooden bird's nesting box, or a cardboard box will work just as well.

Dwarf Campbells Russian hamster.

Food

In common with the other dwarf hamsters, Campbells Russian is not a big eater. If they get too much to eat – and that is very easy to do, because they require a mere 10–15g (⅓–½oz) – they will quickly become too fat. A good mixed hamster food containing all the necessary nutrients is ideal for this type of hamster.

To supplement this diet, the hamster can occasionally be given small amounts of fruit and vegetable such as carrot, apple, or such like that is not high in moisture content. Hay must be available to the animals at all times because they need the fibre.
In common with other dwarf hamsters, the Campbells Russian likes to have some animal protein now and then in the form of dried cat or dog food, or an insect feed for birds.

Care

It is usually sufficient to clean out the toilet corner every two or three days, depending on the size of the cage and the number of hamsters in it. Urine is usually excreted in one, sometimes two places, but the droppings can be found anywhere in the cage. Change the material on the bottom of the cage once each week or every two weeks, depending on circumstances.

Make sure that the cage is not in a draughty position and the window sill is also not a suitable place because it will be too cold in the winter and quickly become too hot in summer. Although dwarf hamsters drink very little, their drinking water bottle needs to be rinsed out and refilled every day with fresh water.
Since these hamsters too store up food in their home, check frequently for food remnants to prevent them from rotting.

Handling

A dwarf Campbells Russian hamster should be picked up in the same way as the other dwarf hamsters.
Many, but not all of them, do not like being picked up and can react strongly against it if you try. If you need to remove and unwilling

Campbells Russian from its cage, put a box with some tasty tit-bits in its cage. The hamster is almost certain to inspect the box and you can move it to another cage.

Sexing

The difference between male and female of this species can be easily ascertained by the distance between the anus and sexual orifice, which are closer together with females than males.

Reproduction

Puberty occurs at about one and a half months for dwarf Campbells Russian hamsters and the females are sufficiently developed at three months to carry, give birth to and suckle a litter of offspring. The female's cycle enables her to be mated once every four days, usually as darkness falls. The gestation period is about eighteen days after which the young are born without hair and with their eyes closed. They weigh about 2g ($^1/_{14}$oz).

The average size of litter is five. At about two weeks, they grow hair and their eyes open, after which they start to explore the surroundings of the nest and to nibble some food with their parents. From three weeks, the young are no longer suckled and by four weeks, they can be separated from their parents. The male shares the rearing of the children with the female so the male should be left with the female.

Life expectancy

Dwarf Campbells Russian hamsters live to between two and three years old.

Special remarks

The two species of dwarf Russian hamsters can be cross-bred with each other, although some of the resulting offspring will not be able to breed. This type of breeding is generally ill advised.

Colours of Dwarf Campbells Russian hamsters

Natural

The natural colours of these hamsters in the wild are brownish grey with dark grey undercolour and dorsal stripe, with an ivory belly. There are three buff-coloured arches on the side that mark the boundary between the colours of the belly and the back. The eyes are black.

Opal

The coat of the opal closely resembles the natural colour of the Campbells Russian,

Natural-coloured Campbells Russian hamster.

Natural-coloured Campbells Russian hamster

Natural-coloured Campbells Russian hamster.

except the black pigments are reduced to a blue/grey tint. The eyes are dark

Argente

Argente is an attractive variety with a rich ginger coat, blue/grey undercolour, ivory belly, chocolate brown/grey dorsal stripe, and ivory belly. The eyes are pink.

Albino

Albino specimens are totally white, without any pigment whatever.
The eyes are pink.

Campbells Russian hamsters: albino (left) and natural.

Mottled

Mottled specimens can be any of the colours, such as natural, argente, or opal. The mottling is not restricted to one particular part of the body and some animals are more pronouncedly mottled than another. Cross-breeding between mottled specimens can cause problems, with some of the offspring being born without eyes or with very small ones. Many of these animals do not live to maturity.

Normal coat dwarf Campbells Russian hamsters

Origins

The normal-haired coat is the original form for dwarf Campbells Russian hamsters. The satin and wavy coat types are mutations of the standard coat.

External characteristics

Campbells Russian hamsters are one of the

Satin-haired Campbells Russian hamster.

short-tailed group of hamsters. Their bodies are about 70–90mm (2¾–3½in) long and they weigh about 40g (1½oz). The tail, which has little hair on it, is about 10mm (⅜in) long but is difficult to see. The males are generally slightly bigger and heavier than the females.

Coat

The normal coat of these hamsters is short and dense, laying smoothly against the body. The pads are also covered with hair.

Colours

The normal-haired type is found in the full range of colours of dwarf Campbells Russian hamsters.

Satin coat dwarf Campbells Russian hamsters

Origins

Satin-coated dwarf Campbells Russian hamsters originate in the United Kingdom where the first example with a different type

Satin-haired Campbells Russian hamster.

of coat was discovered in 1981 among a litter of normal-coated Campbells Russian hamsters.

External characteristics

The same judging requirements apply to examples with satin coats as the normal-haired types.

Coat

The main characteristic of satin-haired Campbells Russian hamsters is the high lustre of the coat. Satin coats occur with each of the colours but this variety is not widely popular because the gene which causes the stain coat – which gives other hamsters a wonderful sheen – causes these hamsters to appear to have greasy, unkempt hair.

Colours

Satin coat dwarf Campbells Russian hamsters are bred in the full range of colours of this particular hamster.

Chinese hamster (*Cricetulus griseus*)

Origins

The Chinese hamster is not part of the short-tailed group of hamsters, belonging instead to the longer-tailed group.
The length of tail is an external difference, but there are differences in social behaviour between the short-tailed and long-tailed hamsters.
Chinese hamsters chiefly originate from the north of China where their natural habitat is diverse: living both on the extensive and barren plains and in dense forest. The first reports of Chinese hamsters date to about 1900 at which time they were treated as Daurian hamsters (*Cricetulus barabensis*)

It was not until the 1960s that they were kept by enthusiasts and started their path to popularity.

Behaviour

Chinese hamsters are virtually only active in the twilight and night-time. During the day they tend to retreat to their sheltered sleeping place and are rarely seen. Chinese hamsters are excellent climbers and pretty lively.

One of the disadvantages of these animals is that they are not particularly friendly towards their own kind. They have to live on their own just as Syrian hamsters do.
Although the animals are quite intolerant and aggressive, they are friendly towards their handler. They only bite if they feel threatened and that is fortunately not often. Chinese hamsters that are calmly approached should not bite their handler. They are easy to tame and it is possible to quickly win the confidence of even adult specimens.

Accommodation

Chinese hamsters can be kept either in a conventional cage with bars or in a glass "aquarium" cage. Each of these options has its advantages and disadvantages. A cage with bars provides an opportunity for the hamster to climb, provided the bars are

Chinese hamster

Mottled Chinese hamster

which had been written about prior to this time. The Daurian hamster looks much like the Chinese hamster but in contrast with the Chinese hamster, it has not become a popular pet. Chinese hamsters were originally kept in laboratories and universities.

horizontal. If the choice is for glass, it is important to provide something on which a hamster can climb to get exercise. One advantage of glass cages is that no dirt falls out to mess the surrounding area.

Scatter a good layer of sawdust on the bottom of the cage but it is best to avoid sawdust from pine because its strongly aromatic smell appears to cause problems for these animals. It is also good to buy an exercise wheel or tread-wheel, and a flower-pot, hamster house, cardboard box, or bird's

Chinese hamsters use their tails to help them climb.

Chinese hamsters

Male and female Chinese hamsters

nest is needed to provide a sheltered place for the hamster to sleep. Hay is an essential, which hamsters both eat and use to line their nests.

Food

Chinese hamsters do not eat very much, in common with the other dwarf hamsters. They do not tend to getting overweight because of their active life style. Give them a well-balanced hamster food, supplemented occasionally with a small piece of a fruit or vegetable. Make sure they do not get much green vegetable, because this can cause diarrhoea.

Most Chinese hamsters like to have some animal protein now and then in the form of dried cat food or a mealworm. They eat quite a proportion of animal protein in the wild and certainly appreciate it in their diet. Hay is important to the diet for the fibre it provides and a Chinese hamster cannot eat too much of it. In the wild, these hamsters drink very little but it is important to ensure that they can drink whenever they wish to.

Care

Chinese hamsters have virtually no smell. They take care of their own grooming and therefore do not need any intervention on your part. The cage should be cleaned out about once each week and thoroughly disinfected once per month. Chinese hamsters

Chinese hamsters with offspring.

deposit their droppings throughout the cage, in common with most hamsters, but these are fairly dry so that they rarely give rise to infections.

They always urinate in the same corner so that it is sensible to clean this "toilet corner" more frequently – about every two to three days. Chinese hamsters store up food in their "nest" so it is wise to check regularly that there is no rotting food.

Handling

Chinese hamsters are not animals to live in groups but live on their own. The females in particular can react aggressively to their own kind.
Although these animals are quite intolerant and aggressive, they are exemplary towards their handler. An amusing trick of these hamsters is to hold on to a finger using both their legs and flexible tail.

Sexing

It is easier to ascertain the sex of Chinese hamsters than with almost any other hamster. Seen from above, the bodies of males finish in a pointed rump, while the females are rounded.
The pointed hindquarters with males is due to a relatively large scrotum. The difference can be most easily determined with young specimens by comparing the distance between the anus and sexual orifices. These are closer together with females than with males.

Reproduction

Puberty with Chinese hamsters occurs at four weeks but the females are not really ready to
successfully rear young until she is two months old. The female is able to mate

once every four days and indicates this by standing dead still and lifting her tail if she is approached by a male.

The female Chinese hamster is always put in with the male and not the other way round. It is also possible to let them get to know each other, if necessary, on neutral territory.
If the female reacts aggressively, the animals must be immediately separated, because fighting can become serious. After the female has been covered, return her to her own cage. The gestation period is about twenty-one days, after which six or more young are born. The offspring are very small and enter the world without hair and their eyes closed. They will be ready to explore outside the next at two weeks and once they are four weeks old, they can in principle be parted from their mother but another couple of weeks does no harm.

Although these hamsters reach puberty at about four weeks old, it is unusual for

them to be mated under three or four months. During the period that the mother hamster is suckling her young, she is very susceptible to stress and it is best to leave her in peace at this time.

Life expectancy

The life expectancy of Chinese hamsters is between two and four years. They therefore have substantially longer lives than short-tailed hamsters.

External characteristics

Chinese hamsters have a slender body that measures about 100mm (4in) from the point of the nose to the root of the tail, and they weigh about 40g (1½oz). The tail can be clearly seen with these hamsters, partly because their hair is shorter than other dwarf hamsters. The tail is about 20mm

Mottled Chinese hamster

(¾in) long. The Chinese hamster uses its tail as an aid when climbing.

Coat

Chinese hamsters have coats of short hair with a sheen that lays smoothly against the body.

Chinese hamster colours

Natural

The natural colour is the original colour for these hamsters. It is brown with black ticking and a very dark brown to black dorsal stripe. The belly is ivory with a dark undercolour.

Dominant spot

The first recorded mutations with Chinese hamsters were noted in the United Kingdom in 1981.
These animals have grey/brown bodies and plain white bellies without an undercolour and white patches on the body, that can vary from one or two to many from one specimen to the next.

The gene responsible for this marking is dominant yet breeding these types is still not easy. Breeding with a dominant spot produces smaller litters because some of them die in the womb.

At shows, judges look for a white patch on the head in addition to those on the body.

White

The first white Chinese hamsters were bred in Switzerland, where it is believed that the responsible gene is dominant.

Since entirely white hamsters sometimes can be born to spotted parents that are recessive for white, it is quite possible that there are two different types of white gene for these hamsters – a dominant one, and also a recessive one.

These white hamsters are a fine all-over white, except for a dark dorsal stripe and dark eyes. This white colour is not yet universally available. A disadvantage of this type is that the males are infertile.

Natural coloured Chinese hamster.

6. Squirrels

Siberian chipmunk or ground squirrel

(Eutamias sibiricus)

History

The Siberian chipmunk is a ground squirrel that is found in the wild in the northern parts of Asia and Russia. Unlike other species of squirrel, this creature prefers to remain on the ground instead of living in trees.

Behaviour

The Siberian chipmunk is diurnal as opposed to nocturnal. These are lively, inquisitive creatures that are busy almost the entire day collecting food and storing it in their burrow. The burrow is below ground and can be more than a metre (yard) deep. These animals like to jump and do so often. In their natural surroundings, they hibernate in winter but do not need to do so in a centrally-heated house. These animals do not need to hibernate for any health reason.

The Siberian chipmunk is a busy little animal that cannot be allowed to wander in the house unsupervised and is not one for a lot of cuddling and stroking but they are very interesting to look at. If young animals are regularly fed by hand, they will quickly become tame but never animals to fondle. Older Siberian chipmunks, which have had little contact with people, rarely become tame.
Siberian chipmunks are solitary animals that only share each other's company when mating. Not too much can therefore be expected of these animals if they are based in the same cage. They can be so intolerant of each other that they will fight to the death,

Left: a Siberian chipmunk.

Siberian chipmunk in a nesting box.

especially if the cage is small. In the wild, they live in each other's presence but each have their own territory that is heavily defended against interloping Siberian chipmunks.

Accommodation

Siberian chipmunks can be kept either indoors or outside, for instance in a specially prepared aviary. Those animals that are kept out of doors will hibernate when the weather gets cold.
The accommodation must provide somewhere for the animal to shelter that is certainly free from frost. Prior to hibernating, they store up food in their burrow. These

could get their legs caught. River dredged sand is the best choice for the floor of the cage, including an indoor one. Cat litter and wood shavings will often be eaten and this is not good for the animal's health.

If you wish to keep more of these creatures together, then the accommodation will need to be at least twice as large. In such an event, make sure there are more besting boxes. These creatures can manage to live together but things can sometimes go wrong. Make sure when the animals are first put together that you are on hand to deal with any problems that arise.

Food

Siberian chipmunks squirrel away their food. If the feeding bowl is empty, that does not necessarily mean that the animal has eaten it all. It has probably taken the food to its burrow. For this reason try to avoid giving them food that can quickly rot or ferment, and then only in small quantities. There is not yet any ready-made food for these animals so you will have to mix it yourself. You can give them a mixture of bird seed and rodent food. The diet needs to be varied, not liable to rot, and certainly low in fat content.

They love to eat sunflower seeds but too much of these can make them grow fat. These creatures eat animal protein as well as vegetable matter and this can be given in the form of mealworms, crickets and grasshoppers, or a packed insect feed for birds. You can also give them dried cat food from time to time. Siberian chipmunks like to use their teeth and must be given things to gnaw on. Branches from willows or fruit trees are ideal, or you can give them a chewing block specially for rodents. If you are uncertain whether your chipmunk is getting adequate minerals in its diet, you can also give it a mineral lick. Fresh water must be available of course at all times, preferably in a drinking bottle because these creatures are likely to knock over a bowl.

chipmunks will burrow into the ground if they are out of doors so there must be a barrier, such as a concrete floor, to prevent them from escaping. Although in the wild they burrow below ground, these animals will happily utilise a nesting box that is filled with hay and fixed off the ground. One animal will often use more than one nesting box.

Despite their compact size, Siberian chipmunks need fairly spacious accommodation. Whether indoors or out, a cage needs to have a floor area of not less than 500 x 500mm (20 x 20in) and not less than 1.5m (5ft) high. The height is specially important because of the way these animals jump and climb. If they are kept in too small an area, with too little to occupy them, these animals can suffer from severe behavioural problems.

This is best avoided by ensuring there are plenty of branches, stumps, and other such things for them to climb on in their accommodation. The material from which the cage is made needs to be rodent-resistant and without any sharp protrusions or bars which get closer together in which these animals

Care

Siberian chipmunks are clean animals that often deposit their excreta in just one place.

This part of the accommodation must be cleaned regularly and disinfected. Check at frequent intervals in the nesting boxes for any food remains that could be rotting or fermenting. These animals do not need any grooming because they keep themselves clean.

Handling

Siberian chipmunks must never be picked up by their tails because this can very easily break off and will not grow back.

Sexing

There are no obvious external differences between the two sexes. The distance between the anal and sexual orifices is greater with males than with females.

Prevost squirrel

Both the males and females have virtually no smell.

Reproduction

Siberian chipmunks reach puberty at about nine months old. The females announce with a whistling sound when they are prepared to mate. This usually occurs every two weeks and lasts for about two to three days.

A female that issues this sound will normally not attack a male that is placed in her cage but once the pairing is completed, the female cannot bear the male any more in her presence and he needs to be removed for his own safety. This is quite different when a pair of Siberian chipmunks live together. The male will withdraw to his own territory after mating.

The pregnancy lasts about thirty-one to thirty-two days and some three to five young are born. These are both without any hair and have their eyes closed. Once they are about one month old, they leave the nest to get to know their surroundings. Once they are two months old, they can look after themselves and be separated from their mother.

Life expectancy

The average life expectancy for Siberian chipmunks is ten years.

External characteristics

Siberian chipmunks are about 140–170mm (5½–6¾in) long from their nose to the root of their tail and they weigh on average about 100g (3½oz). The tail, which is covered with hair, is about 100mm (4in) long. These animals have characteristic cheek pouches in which they can collect food.

Coat

The body of the Siberian chipmunk is covered with short, shiny hairs that are soft and lay sleekly.

Siberian chipmunks have a greyish to golden-brown coat with five striking dark stripes running from their neck to the root of their tails. The belly is always white.

Other squirrels

There are many different species of squirrel world-wide, divided into about forty different families. The representatives of this large family are very widely varied. The majority of them are active during the day and sleep at night.

The largest of them are the marmots (Marmota), which weigh at least 7kg (15½lb) and the smallest are the dwarf squirrels weighing a mere 10g (1/3oz).

The characteristic feature for most squirrels is their bushy tail which they use as a "paddle" when they jump from branch to branch and they also wrap around themselves to protect them from the cold. There are two main types of squirrel: those that live in trees and the ground squirrels.

The ground squirrels include the popular Siberian chipmunk, the American chipmunk with its thirteen stripes (*Citellus tridecimlineatus*), the prairie dog of the genus Cynomys and the Eastern chipmunk (*Tamias striatus*).

The true tree-living squirrels include the colourful prevost (*Prevost borneo*) and the various Oriental giant flying squirrels (*Petaurista*). Flying squirrels, unlike most other squirrels, are active in the twilight. Their name is derived from the flaps of skin between the front and back legs which is used to glide from tree to tree.

Some squirrels are strictly solitary animals which only pair up to mate, while others pair up for life, while some live in groups. Their diet consists largely of seeds, fruits, nuts, and other vegetable matter but most squirrels also like to eat animal protein in the form of bird's eggs, insects, and even fledgling birds.

Prevost squirrel

7. Degus

Degu (*Octodon degus*)

History

Degus originate from the Andes mountains of Chile, South America where these friendly animals live in large groups on the rocks. The first Europeans to discover degus did so in the eighteenth century when they were considered to be squirrel-like creatures and were treated as such by scientists.

Subsequently it was discovered that these were not a form of squirrel but a close relative of that other South American creature: the cavy or guinea pig. Shortly after they were discovered by Europeans, the first specimens were shipped back to zoos in Europe. In the middle of the twentieth century, the majority of captive degus found their way to laboratories, where they were kept to study their social behaviour, and for medical research.

That the degu makes an interesting and companionable household pet has only recently been discovered but with their inquisitive nature, friendly, and often amusing ways, their popularity is increasing.

Behaviour

Degus live in social groupings in the wild, where they scramble and climb over rocks for which they are well equipped. Just like cavies, degus make whistling and growling sounds to communicate and in contrast with many other rodents, these creatures are very much day-time animals, which retreat to their burrows at sunset and come above ground once more with the first rays of the sun in the morning to search for food. They are usually somewhat less active in the middle of the day and very active in the late afternoon.

Unfortunately, degus are often kept on their own but it is far better for them to live in groups, or at very least to have at least one other degu for company.

Degus quickly get used to their handler.

If you really must keep a degu on its own, you will have to spend at least an hour every day giving it your full attention to prevent it developing behavioural problems. Making up a group of degus has to be done with some forethought. It is true that they never fight with each other in the wild but there are sufficient means of escape which are not available in a cage. Female

degus usually get on with each other without problems, even when they do not know each other. The problems are more likely to occur with sexually mature males.

Adult males that did not grow-up together can rarely live together, and certainly never when there are females close at hand.

Provided there are sufficient females and plenty of space, males will tolerate each

Degus only eat vegetable food.

other. Problems usually occur when there is a shortage of females, combined with too little space.

Degus are very active animals and very inquisitive about everything that goes on around them. They rarely bite and quickly establish a trusting bond with their handler.

Accommodation

Degus are best kept in a large glass cage covered with strong mesh. Some degus gnaw so actively that they bite their way through even hard plastic cages specially designed for rodents, so bear this in mind.

The bottom of such a cage or aquarium can be covered with wood shavings but these must be free from dust because degus can suffer from pneumonia if there is too much dust in the scatter material of their cage. Wood shavings that come from pine trees can also be harmful for their health.

Degus like to climb and it is important to provide them with the opportunity by arranging branches and other things for them to climb on in an interesting way. Make sure the cage is not too low – one of 400–500mm (16–20in) is sufficient. Degus, in common with all rodents, need to have things on which to gnaw to prevent their teeth from becoming too long.

Willow or fruit tree branches are ideal but degus will use their teeth on virtually everything you place in their cage. Place a heavy earthenware bowl filled with clean silver sand for the degus to bathe in so they can keep their coats in good condition. Every degu cage needs to also have a nesting box for the animals to shelter in when sleeping or taking a rest.

Food

The natural habitat for degus is rocky terrain with a lot of shrub-like growth. They eat grain, grasses, seeds, and almost any edible plant matter.
There are special food mixes for degus but these are not universally available. If you are unable to find this mixed feed in your area, then you can feed them with guinea pig or rodent food. Degus also like to eat small amounts of fruit and vegetables but make sure you do not give them too much, since this can cause digestive disorders.

Degus that are kept as pets are often given too many tit-bits that are high in calories and low in fibre, which leads to them putting on weight. Becoming overweight can often lead

Degu

Degu

This is best placed in a sturdy earthenware receptacle which can be bought from pets shops.

Care

Degus are fairly house-trained animals which need very little attention. Provided they have a sand bath and plenty to gnaw on, they will take care of themselves and keep themselves clean.
The cage is cleaned out as necessary. This will be more frequently with a small cage housing many animals than in a spacious one with just one or two.

Handling

to diabetes, which is an incurable illness to which degus are very susceptible. For this reason, do not spoil a degu too greatly since this can be unhealthy for them. Hay is an ideal food for degus and there should always be plenty in their accommodation but make sure that it is free of dust.
Water is best provided by means of a glass drinking bottle. Plastic bottles will normally be wrecked by degus. Degus bathe themselves by regularly turning themselves over in clean white sand (not builder's sand).

Its is very easy to tame a degu. The animals are naturally inquisitive and their curiosity usually wins out against caution.
If you regularly feed a degu by hand, it will quickly become used to you. Degus are not true cuddly pets although you can usually

pick them up and stroke them but they do not become as tame and affectionate as guinea-pigs because they are far too active for that. Never pick a degu up by its tail because this may snap off and it will not grow back.

Sexing

The sex of young specimens can be determined by checking the distance between the anus and sexual orifice, which is much greater with males than females.
With sexually mature males, the scrotum is also clearly visible.

Reproduction

Female degus are ready to breed when they are about three months old but they are not ready at that age to rear a litter of offspring. It is better to wait until they are about five months old. The times when a female is ready to mate differs from female to female

Degu

Degu

Degu mother with her young.

but it is usually about every two or three weeks. If a sexually mature male and female are kept together, they will sort out for themselves when the time is right. The male can be left with the female after she has been covered and it is not necessary to put him in a separate cage when the young are born. The pregnancy is relatively long, lasting for about ninety days.

The average size of litter is five but can range from three to even ten. The young are born almost fully formed with their coat, eyes open, and they can run around immediately. At two weeks, they start to nibble some of their parents food but still need their mother's milk.

Once they reach five to six weeks, they are weaned off suckling but are still too young to fend for themselves. It is best to keep the offspring by the mother until they are about eight weeks old.

Life expectancy

Degus normally live to be between five and eight years old.

External characteristics

The degus that can be found in the wild in Chile are often much larger than those which are kept as pets. This is surprising since this is entirely the opposite of the experience with hamsters.

The fact that most animals kept as pets grow larger than their counterparts living wild is due to their getting plenty to eat and less exercise.
The exception with degus is probably due to inbreeding: the present breeding line of pet degus are virtually all related.
In nature, degus grow to about 400mm (16in), including their tail, weighing about 200–300g (7–10½oz).

Degus are very inquisitive.

Those in captivity are almost always smaller and lighter too.

Coat

A degus coat is short but its density depends on its living conditions. A degu that lives in a colder region will have a thicker coat than one in a warmer climate or kept indoors in a centrally-heated home.

Colour

At present, Degus have only been bred in their original natural colour which is brown/grey with black ticking on the ends of the hairs. The domesticated degus can have the odd spot of white on them. Their eyes are dark brown. The tail, which is slightly shorter than the body, is covered with hair.

8. Chinchillas

Chinchilla *(Chinchilla laniger)*

History

The chinchilla is related to the cavy or guinea-pig and originates from South America. There has been interest in these animals for a very long time. The species was discovered by the Spanish conquistadors, who conquered most of South America, in the sixteenth century.

Chinchillas were written about by Molina towards the end of the eighteenth century. In South America, the creatures were hunted for their pelts by both the Spanish and the native inhabitants. Chinchilla fur was regarded as very soft and incomparable with other types of fur then available.

Several hundred years later, Europeans also discovered the wonderful softness of chinchilla and their pelts were keenly traded. The demand was so great the wild population of chinchillas was threatened with extinction. This was the beginning of the twentieth century. The fur trade was so concerned that the chinchilla would disappear and with it the money to be made, that they decided to capture some specimens in order to breed them. Because chinchillas have some very specific needs for their care, there must have been a few difficulties to be overcome, but chinchillas were soon being bred on a large scale in both North America and Europe.

Fortunately times have changed for both the chinchilla and other animals coveted for their pelts. People in the west started to develop an aversion to breeding animals for the fur trade and public opinion caused an enormous drop in demand for fur so that the chinchilla farms saw their turnover drop before their eyes but the world goes on and now many farms breed the animals as interesting household pets.

The first chinchillas that were kept as pets were very susceptible to stress and needed a lot of calm, being unable to cope with noise. Since then, the animals have adapted far better to life in the home.

Behaviour

Chinchillas are very much twilight animals. They creep away in their sleeping quarters by day and only become active during the evening. Some of these creatures do put in an appearance during the day – especially if there are unusual sounds or other things to arouse their interest, so that they come to have a look. Chinchillas are not really pets for smaller children. This is partly because

Left: silver-white chinchilla.

Chinchillas are inquisitive animals.

Chinchillas in the natural colour.

155

Chinchillas enjoy a nesting box.

Chinchillas enjoy a nesting box.

their nocturnal activity is directly opposed to the natural life rhythms of young children, but also because keeping and caring for a chinchilla is no child's play. Looking after chinchillas is nowhere near as simple as, for example, caring for their close relative, the guinea-pig.

Chinchillas are above all gentle creatures and very bright. Once you have gained their trust, they will let you handle and stroke them. They rarely bite because aggression is not part of their nature. Chinchillas are sociable creatures and keeping one on its own is not kind, for the animal will become listless and lethargic, or develop other behavioural problems. Chinchillas that are put together when young can often live their whole lives together without squabbling. Adult chinchillas do not accept each other so easily, although it is not impossible to get older animals accustomed to living in proximity with each other.

Accommodation

The accommodation for a pair of chinchillas needs to be big enough: generally, a cage about 1m (39in) high, with a floor area of about 500 x 500mm (20 x 20in) is usually sufficient. A bird cage, obtainable from better pet shops and specialists, is quite suitable. An additional advantage is that these are entirely made of metal and are therefore resistant to the sharp teeth of the chinchilla. A parrot's cage is not suitable because there is too much room between the bars and they often have a plastic under-tray which a chinchilla would gnaw through in no time at all.

A chinchilla cage should contain a few thick branches for them to climb on. Branches from beech, willow, or fruit trees is ideal. Chinchillas also need a few different places to sit at different heights. It is not essential to have a nesting box or somewhere for the animals to hide away during the day but many chinchillas do use them if provided.

In common with other rodents, the cage should be sited away from any draughts and not where it is in the direct, full sun. Cold parts of the house, or those which are constantly busy are not suitable places. Chinchillas need to be able to "bathe" themselves at least once per day in a roomy tray of sand in order to keep their coat clean. Special chinchilla sand is sold for this purpose, which is finer than normal white sand.

The sand should be placed in a sturdy earthenware dish on the bottom of the cage. The bottom of the cage and the sleeping place should be covered with a generous layer of wood shavings. Water is best given in glass bottles because plastic ones are unable to resist a chinchilla's sharp teeth.

Food

Chinchillas have a sensitive digestive system that is adapted to a high fibre but otherwise rather frugal diet. Too much fat, protein, and other nutrients, or the slightest bit of mould on their food inevitably causes diarrhoea, which in many cases lead to death.
Vegetables and fruit should only be given very sparingly and preferably limited to no

Chinchilla food should be low in nutrients and high in fibre.

Ebony chinchilla

Chinchilla

It is best to buy a specially mixed food for chinchillas. Changing over from one make of food to another can even cause diarrhoea so try to avoid doing so. Chinchillas have great need of quality hay and a chewing block.

Care

Chinchillas keep themselves extremely clean but they need a daily sand bath to do so. The sand can be left in the cage but in view of the relatively high cost of the special chinchilla sand, it is better to place it in the cage each day to prevent it from becoming fouled.

more than one small piece per week and only those that are not high in moisture content, such as a small piece of apple, or a blackberry.

Never give them sunflower seeds, peanuts, lettuce, or any type of cabbage or brassica! Chinchillas have entirely different nutritional needs than rabbits and guinea-pigs, so rodent and rabbit feeds are not suitable.

Caring for chinchillas means inspecting their excreta daily. Normal chinchilla droppings are quite hard but if these are soft then something is wrong. The frequency with which you clean the cage depends on the size of the group of chinchillas and surface area of the cage. Normally, weekly replacement of the scattered material on the floor is sufficient.

Handling

A tame chinchilla will let you pick it up with-out problems but if you have to capture it, it is best to first catch hold of it firmly by the root of the tail while your other hand supports the animal.

Sexing

The difference between the sexes is readily seen by comparing the distance between the anus and sexual orifice. This is farther apart with males than females.

Reproduction

Puberty occurs with a female chinchilla at about five to six months, but it is better to wait until she is eight or nine months old before breeding her.

The female is ready to mate every thirty days or so – there can be some variation of the cycle. She only comes into heat for a few days. If you keep a pair of chinchillas the female will ensure that she mates but if you keep two females together you may after some time have to add a male.

Theses animals do not always readily take to each other and it is best to keep the male in a nearby cage for a few days so that they can get used to each other. If they react positively to each other, you can put the male in the cage with the female.

The male can remain with the female while she suckles her young.

The pregnancy lasts for about 111 days and litters usually comprise one to three offspring. These are born fully formed, with hair, eyes open, and able to walk. They weigh about 40–55g (1½–2oz). They are sufficiently independent by eight weeks to fend for themselves without their mother.

Chinchilla

Silver-blue chinchilla

Young chinchilla

Chinchilla

Two-colour spotted chinchilla

Life expectancy

A well-cared for chinchilla can live to quite an age. The average life expectancy is about fifteen years.

External characteristics

Chinchillas are about 300mm (12in) long from nose to the root of the tail. The tail is covered with hair and measures about 150mm (6in).

The weight of a chinchilla is between 450–700g (15¾oz–1½lb) with the males being the larger specimens.
The head is neither large nor small but the ears are relatively large and oval, and stand

erect. The expressive eyes are almond shaped to rounded, giving a friendly and intelligent appearance.

The whiskers are unusually long, sometimes 100mm (4in) long but they can be up to 150mm (6in) long.

Coat

Chinchillas have a silken coat that is extremely densely formed. The hairs do not lay flat against the body but stand up.

Colours

Chinchillas originally only had the one colour which has given it name to a similar rabbit colour. Recently further colours have been developed, such as brown, beige, blue velvet, various shades of white (with which the colours of the eye can vary), and multi-coloured chinchillas.

It is significant that breeding between white chinchillas and velvets can be wholly unsuccessful or produce offspring that are unsound. These colours should therefore never be bred with each other.

9. Cavies or guinea-pigs (Cavidae)

Cavies (*Cavia porcellus*)

History

Guinea-pigs, or cavies as many enthusiasts call them, originate from South America where they were more or less kept as pets by the Incas, long before Europeans conquered that continent. When the Spanish conquistadors discovered these creatures at the beginning of the sixteenth century, they found them around the people's houses, fairly well domesticated. In addition to the natural golden agouti colour there were other colours too and the animals were quite tame. It is likely that the Incas kept their guinea-pigs for consumption. The animal is still considered a delicacy in South America, where it can be found on the menu in these countries. The natives also used guinea-pigs as offerings to their many gods. It is not known whether the Incas kept guinea-pigs as pets but considering how children the world over are interested in animals, it is highly probable that children played with them.

It is not entirely certain when guinea-pigs arrived in Europe because reports contradict each other. Some suggest that they were brought back by the Spanish as early as the sixteenth century but others suggest that English seafarers introduced them to Europe later than this. One thing that is certain is that they were being kept in a number of European countries at the beginning of the eighteenth century. Europeans did not have much regard for the guinea-pig as meat because they had far better alternatives with their pigs and cattle that were unknown in South America.

The majority of guinea-pigs came at first to laboratories but they soon become popular pets for children. Because these animals were considered exotic, unusual, and from distant shores, they were very expensive and therefore restricted to the wealthy. They were also put on show as curiosities at fêtes and fairs. The breeding of guinea-pigs became very much a British affair and shows were held in the Britain in the nineteenth century at which guinea-pigs were judged on their appearance. The English name guinea-pig was widely used outside Britain too for a long time but the named cavy, derived from the Latin Cavia, is now used among British enthusiasts for small rodents alongside the popular common name.

Guinea-pigs are now the most widely held and best loved of pet rodents and their diverse original colours have been further extended in range. There have also been additions of different types of coat.

Left: tri-coloured cavy.

Suckling mother cavy with two day old young.

Golden agouti is the original cavy colour,

Cavies naturally live in groups.

Texel cavy

Behaviour

DCavies are sociable animals that are active by day and ideally suited as companionable pets. In the wild, these creatures live in large colonies and when kept as pets they like to have the company of other cavies or at least sufficient attention from their handler.

If you wish to keep a group of cavies together, it is better to have sows than boars because adult boars can be very intolerant of each other and if their are sows around, they are liable to fight each other.
Cavies make a variety of sounds with which they communicate: they can make it very obvious that they want more attention, have been frightened, or are hungry. Cavies are renowned for virtually never biting. If they are afraid, they do not defend themselves by biting but freeze rigid. Cavies are very clean animals that clean themselves several times each day.

Accommodation

Cavies can live out of doors in the summer, perhaps in a rabbit run. The run does not necessarily need a roof to keep the cavies in because they do not jump but they do need protection from cats, foxes, and dogs. The sleeping hutch must be fully draught-free. Cavies are not very resistant to cold, heat, or rain, and need protection from these elements.

When the temperature falls, they must be moved to a draught- and frost-free shed or outhouse. They cannot cope with temperatures that are too warm or that constantly change so you must never move them straight from an outdoors hutch to a heated room or vice versa.

The ideal ambient temperature for cavies is 17°C–24°C (62°F–75°F). Ideal outside accommodation for cavies has a hutch to provide shelter from the rain, sun, and cold winds. Make sure they cannot escape from their run by burrowing their way out. They are quite happy not to be burrow in their indoor cages, and it does not them no harm to deny them the chance when out of doors.

Cavies that live out of doors do not become as tame as those that are constantly in contact with the members of the family. An ideal indoor hutch should have a floor area of at least 600 x 400mm (24 x 16in) and preferably larger if the cavies do not get the chance to stretch their legs outside it at regular intervals.
If a cavy gets too little exercise, it will quickly put on weight which will shorten its life. Cavies like somewhere to shelter, even in an indoor hutch or cage. Such cages can be bought from a pet shop but it is easy to make one from substantial timber. The bottom of the cage should be covered with wood shavings but make sure these are not sharp. Straw is unsuitable because it too hard and sharp but also because it does not absorb much moisture. The feeding trays or bowls need to be substantial to prevent the cavies from chewing them to pieces or tipping them over. Water is best provided from a drinking bottle placed outside the cage.

Food

Cavies have two important dietary requirements to remain healthy. The first is that they must have plenty of raw fibre. The best fibre provider for cavies is hay, and sufficient must always be available to them in their cage. Put the hay in a net rather than laying it on the floor, where it will quickly become dirty.

In addition to raw fibre, cavies need daily intake of vitamin C, which they cannot produce themselves. A cavy that receives too little vitamin C will quickly become ill and die, so do not underestimate its importance. The vitamin C can be given in the form of tablets, or drops in the drinking water, but you can also ensure a cavy has some fresh fruit and vegetable each day. Some brands of cavy food have the necessary vitamins added but this is not always the case.

Requirements can vary but on average cavies eat about 60g (approx. 2oz) food per day. Apples, chicory, carrots, kale, and endives are suitable fruit and vegetables for feeding to cavies. Cabbages and lettuce are best given in very small amounts and then only to adult animals because cavies have a very sensitive digestive system and these foods can cause diarrhoea. Make sure cavies do not get any green vegetables that are not fresh, mouldy, or rotting.

Basic food can be a ready-to-use cavy food. Do not use cavy foods that contain lots of sunflower seeds, peanuts, green pellets of pressed silage, or other coloured pellets. Do

Smooth-haired albino cavy.

Lilac agouti Texel cavy.

not feed them with rabbit pellets either, because these are too low in nutrition. Give a cavy a clean willow branch that has not been sprayed to gnaw on and they will also enjoy a mineral block hung through the bars of the cage.

Care

Cavies are easy pets to care for, which do not need a great deal of exercise. Exceptions to this are the long-haired cavies, which need daily grooming with brush and comb. The growth of the claws need to be kept in check; they can get so long that they need clipping.

Overlong "elephant's tusks" can occur if the incisors do not wear down properly, in common with most other rodents. If you see that your cavy is not eating properly or finding it difficult to eat, take the animal straight away to a vet or cavy breeder, who can trim the teeth back to size. Once a cavy

Fresh vegetables are an important source of vitamin C.

Most cream cavies are too dark coloured to be shown.

has developed these overgrown "tusks", it will suffer from them for the rest of its life and will need constant attention.

The frequency with which you need to clean out the hutch or cage depends on the number of cavies its houses. One cavy kept in a spacious cage will only need cleaning out once each week but more animals in a small cage will need cleanly more often. It is a good idea to disinfect the cage once each month. The water in the drinking bottle must be replaced every day.

Handling

Cavies are relatively easily tamed, especially if they are hand fed from a young age. If kept in a more distant shed or out of doors without much contact, then they will be tamed less easily.

Always approach a cavy calmly, at eye level. You will win a cavy's trust more quickly if you regularly offer it tasty tit-bits in your hand. A cavy is always picked up by fully supporting it and make sure that it does not fall because it is likely to break something.

Sexing

The two sexes can be differentiated by lightly pressing the sexual orifice. This is best done by holding the cavy on its back, with its head towards your chest.
The boar's sexual organ becomes visible by applying light pressure. With adults, the difference is quite apparent because the boars are noticeably larger than the sows.

Reproduction

Female cavies are ready to mate about every sixteen days. Most sows that are ready to be covered make grunting sounds that can easily be recognised because they differ from the usual sounds that cavies make. Always place the sow in the cage with the boar and never the other way round. Their reaction to each other will make it apparent whether the sow is ready to mate.

The pregnancy lasts for between sixty-five and seventy days. The boar can be left with the sow and the litter but most breeders separate them to prevent the sow from being covered again soon after she has given birth. Carrying the young, giving birth, and suckling the young takes a lot out of the sow and it is best to limit her to two or at most three litters per year. After she has been covered, it is best to put her in a cage on her own where she can calmly prepare for the great event.
During the pregnancy, leave her in peace as much as possible and try to protect her from any stress which can lead to a miscarriage. Try to avoid picking up a pregnant cavy and if it has to be done, make sure no pressure is applied to her belly.

Most cavies give birth easily but problems can occur. Cavy offspring are quite large when they are born, weighing on average about 90g (about 3oz). Cavies that do not

Tri-colour day-old cavies.

have their first litter until they are older than eleven months can have major difficulties in labour.

Cavies above this age have pelvic bones that are less resilient than with younger animals, so that giving birth to such relatively large offspring is very difficult. The best results are obtained with sows that have their first offspring before they are eleven months old but not younger than six months. The ideal age for a first litter is between eight and nine months.

Cavies give birth to two to four young, which are miniatures of their parents at birth: they are fully developed and can walk almost immediately. They are covered with hair, their eyes are open, and they have teeth.
Within a few days, they will nibble at their parent's food but they need their mother's milk until they are about four weeks old. Depending on their development, the young can be separated from their mother from five to six weeks old. Part the boars and sows from each other so that the young sows do not unwittingly become pregnant at too young an age. Many cavies are physically capable of breeding at two to three months but they are not yet fully mature.

Life expectancy

A cavy has a life expectancy, dependant on its health and standard of care, of six to eight years. Some live longer but the majority, especially if extensively bred, reach five or six years old.

Golden agouti cavy

Silver agouti cavy

Cavy colours

Agouti

Golden agouti is the original and natural colour of cavies. Each hair of the agouti has a number of dark and light bands of colour. Some cavies also have coats in which parts of the hair are wholly dark or just the tip.

The finest agouti coats have evenly distributed colour ticking throughout the coat, or in other words, the dark and light bands are equally and regularly distributed.

Show judges regard light rings around the eyes as a fault and white spots or patches are also unpopular on the show bench. The belly is always free from ticking. There are various agouti colours, with the more usual ones being described below.

GOLDEN AGOUTI
Golden agouti is the original and natural colour of cavies. The black agouti ticking appears with a warm reddish-brown base colour. Golden agouti cavies have black claws, ears, and soles to their feet, with dark brown eyes.

GREY AGOUTI
Grey agouti resembles golden agouti but the base colour is paler and tends more to the yellow part of the spectrum than red.
The ears, pads, and claws are also black with this colour, and the eyes too are dark brown.

SILVER AGOUTI

Silver agouti cavies have a black ticking in their coat and a gene that prevents red pigment from being formed on the hairs, so that everything that is red or yellowish with the golden and grey agouti colours is white. These cavies have dark brown eyes and black foot soles and claws.

CINNAMON AGOUTI

With cinnamon agouti, the black pigment of the ticking is mutated to a colour resembling cinnamon. The base colour is ivory, like the belly.

The pads, claws, and ears are all brown, together with the eyes. Cavies that become this colour are born darker but lighten as they get older. Once they have lost their original hair they were born with, the eventual colour is fully developed. This usually happens at about five months of age.

SALMON AGOUTI

Salmon agouti cavies have an ivory under-colour with lilac ticking and an ivory base colour. The ears, which have virtually no pigment, are pink, the eyes are pink, and the claws are natural horn coloured.

Remarkable is that the young are born dark but later acquire their permanent salmon colouring.

Self-coloured cavies

Single plain colour (known as self-coloured) cavies are extremely popular. These cavies

Salmon agouti cavies are born dark; their permanent colour develops when they are five months old.

Smooth-haired self chocolate cavy.

have no ticking in the hairs in their coat and ideally are the same shade of colour over their entire body. Each hair should be one colour from root to tip and their should be no white patches, blazes, spots or other colour at all present in the coat for showing purposes.

BLACK

A black cavy should be pitch black with each hair solid black from tip to root. The feet and claws are also black but the eyes are dark brown.

CHOCOLATE

Chocolate cavies should be as dark brown as possible with brown soles to the feet, claws, and ears. Their eyes are dark brown but in the right light they can glow fiery red.

LILAC

Lilac cavies coats should not be tinged with brown and the preference at shows is for as pale a shade of lilac as possible. An ideal lilac cavy has pads, the feet themselves, and the ears without any pigmentation.

Their eyes are pink. Like other self-coloured cavies, there must not be any other colour present in the hairs of the coat.

Lilac cavies are born dark and only develop their eventual coat when they are five to six months old.

BEIGE

Beige cavies have a dark ivory coat with a tendency towards grey rather than brown being preferred.

The pads and ears are flesh-coloured (without pigment) and the eyes are pink. In common with other self-coloured cavies, there should be no other colour hair in the

Self beige cavia

Smooth-haired self-coloured red cavy

Smooth-haired self-coloured cream cavy

coat. These cavies are also born with dark coats which change to their eventual colour when the animals are five to six months old.

RED

Red cavies are a warm chestnut red. Show judges have a preference for darker shades of this colour rather than lighter ones. The coat should contain no other colour. Red cavies have dark eyes and ears, with black soles of their feet and claws.

GOLDEN

Golden cavies actually have coats of orange. There are both dark-eyed and pink-eyed golden cavies.

The dark-eyed types have black ears and pads while the pink-eyed types are virtually without pigment.

The colour should be similar across the entire body, without nuances. These cavies too are born far darker than their parents and acquire their eventual colour at four to six months old.

BUFF

Buff cavies have a warm yellow ochre coat without any tinge of red. At shows the preference is for ears and pads that are entirely free of pigment. Like other self-coloured cavies, these too should have no markings of another colour. The eyes are dark brown.

CREAM

The preference is for self cream cavies to have a light ivory cream but in reality many cream cavies are far darker.

The same colour and shade should cover the entire coat without any markings. The eyes are brown, with flesh-coloured pads and ears.

ALBINO

White cavies with pink eyes, or albino, should have a pure white coat without any tinge of yellow.

The ears, pads, and claws are without any pigment.

DARK-EYED WHITE

Dark-eyed white cavies have coats that are entirely white and either blue or brown eyes. The rest of the body is free of any pigmentation.

Cavies with markings

Cavies are also bred with a variety of difference markings and colour combinations. Breeding these markings is a challenge for the more experienced breeders because the characteristics do not breed true.

Two perfectly marked parents can have offspring that are poorly marked but the reverse is also possible. Certain of the more usual markings and combinations are described below.

BRINDLE

Brindle cavies have a mixture of black and red hairs in their coat. The preference is for an even distribution and mixture of the two colours. Where a lot of red or black hairs are clustered together, this creates patches of that colour, which are considered a fault.

White hairs or patches are equally frowned upon. The ideal brindle has an even distribution throughout the body so that there are no darker (clusters of black hairs) or lighter (red hairs) patches but in reality this ideal is rarely achieved. The ears and pads should be black with dark eyes.

TORTOISESHELL

Tortoiseshell cavies have coats with even markings of two colours (red and black). The patches should ideally be clearly defined and more or less rectangular, almost as if the line between the colours had been drawn with a ruler. Seen from the side, these animals should have between three to five patches of colour and these should be divided along the back in a straight line.

The opposite colour patch should always be of the contrasting colour so that no banding is formed around the body. Tortoiseshell cavies have dark eyes, ears, and pads.

This very attractive coat pattern is a major challenge for breeders because the ideal standard is extremely difficult to achieve. Breeding two perfect specimens does not assure equally well marked offspring.

This pattern of coat is also bred in other colour combinations.

TRI-COLOURED

Tri-coloured cavies are tortoiseshells with the addition of white.

The challenge with this pattern is to achieve three colour patches on the side that are each of a differ-ent colour. The boundary between the colours is ideally as straight as possible along the centre of the back. It is permissible for tri-coloured cavies to have eyes of different colours.

Tri-coloured cavies are not just black/red/white but also chocolate/red/white.

DUTCH

The Dutch breed of marked cavies have similar markings to Dutch rabbits – the front of the cavy is white and the rear is coloured, with two or three patches of colour on the head, ears, close to the eyes, and on the cheeks. These patches ideally should be symmetrical, meaning that a patch on one side should be balanced by a patch of equal size on the opposite side of the body. The partition between colour and white at the centre of the body should be as equally sharp and straight as with Dutch rabbits.

The feet of the hind legs are white. The coloured parts of the body can be any colour but the most usual Dutch cavies are black.

ROAN
Roan cavies have a mixture of white hairs with different colours, including black, red, and mixed black/red. The white hairs should not be too dominant or visible in patches but evenly distributed throughout the coat across the entire body.

DALMATIAN
Like the dogs of the same name, Dalmatian cavies are white with dark spots which should be evenly distributed across the body and clearly defined, not running in to each other.
The colour is at present not recognised in many countries.

TRUE BREEDING COLOURPOINT
In contrast with the previous markings, there is a type of colour marking, not dissimilar with the colour-pointing of extremities with Siamese cats, which does breed true. With this type of colour marking, two cavies exhibiting colourpoint markings will produce offspring with the same markings.

This is a popular type of cavy, in which the body is white, rather than the creamy-beige of Siamese cats but in common with rabbits with these markings, cavies too have pink eyes. The nose, feet, and ears can be black or chocolate. The muzzle marking should be rounded and not too small and the boundary between the colour and white should be clearly defined. Chocolate is recessive by comparison with black which means that two black colourpoint cavies can produce offspring with chocolate markings but not black from two chocolate parents.

The variety itself is recessive in comparison with all other types of marking. The eyes are always pink while the claws and pads are dark. It is difficult to keep an older cavy with these markings in show condition because the white tends to show dark patches through exposure to sunlight and moisture.

The "shadow patches" this forms are detrimental to success at shows. This type of cavy is born white, with the darker extremities only being fully formed after about five months.

Smooth-haired cavy

External appearance

Smooth-haired cavies should be powerful and muscular. Their bodies are stocky and nicely rounded. A striking characteristic of a cavy is its high, broad, and in every respect well developed shoulders which slope gradually down to the short, broad, and straight back.

The chest is deep, broad, and full. The legs are short, straight, and sturdy. Smooth-

A fine group of colourpoint cavies.

Tri-coloured cavy.

Salmon agouti smooth-haired cavy.

Black cavies should be pitch black.

A group of colourpoint cavies.

haired cavies have four toes on the front feet and three on the feet of the hind legs. Judges pay particular attention to the shape of the rump at shows, which should be well rounded, and well covered with flesh so that no bones are visible.

Cavies have powerful, fairly short necks, with a stubby, broad head. Pointed heads occur regularly with smooth-haired cavies but these have little success at shows.

Viewed from the side, cavies have an arched or Roman nose profile. The ears are relatively widely spread because of the width of the head.
The head has well developed cheeks. Ears are hairless and set at the slant and pointing downwards and with a crease in their middle. The eyes are large, round, and bulbous, with a clear expression.

A well developed adult smooth-haired cavy weighs 900g–1,200g (2lb–2lb 10oz). If you wish to show a cavy, make sure that its claws are kept short, if necessary by clipping them

so that they make a straight line with each other. Cavies have no visible tail.

Coat

Smooth-haired cavies are also known as normal-haired types because this is the original coat type. The short hair is about 30mm (1¼in) long, laying smoothly against the skin. The undercoat is soft, while the outer hairs are coarse. The coat should have a definite gleam. Smooth-haired cavies are the most numerous types.

Colours

Smooth-haired cavies are bred in a wide range of colours, the majority of which are recognised.

Properties

Smooth-haired cavies require virtually no special care to keep their coats in order. In common with other cavies, the claws may need clipping from time to time, which can be done by a vet if you prefer, or by a dog trimming specialist, or experienced breeder.

The smooth-haired types are ideal for showing markings off because the markings always appear more sharply defined on the smooth coat.

Smooth-haired satin beige cavy.

Golden smooth-haired satin-coated cavy.

Satin-coated smooth-haired buff coloured cavy.

Tri-colour chocolate satin rex-coated cavy.

Smooth-haired silver agouti cavy with satin coat.

Satin coated cavies

External characteristics

Satin coated cavies have the same body conformation as normal smooth-haired cavies.

Coat

Satin-coated cavies are smooth-haired. The hairs of the satin coat are, however, softer that the normal coat and the coat has a deep lustre which imparts depth and warmth to the colour. Satin coats should have dense hairs with a lustre deep into the root of every hair. The coat is about 30mm (1¼in) long, and lays sleekly against the body. This type has relatively little undercoat.

Colours

Satin-coated cavies are bred in a range of colours. In principle, every colour should be available with a satin coat but some colours are not yet recognised. Common colours are self red, golden, cream, buff, and white.

Smooth-haired red cavy with satin coat.

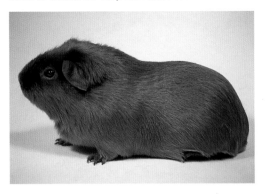

Lilac/golden tortoiseshell English crowned cavy.

There are also agouti cavies with satin coats.

Properties

Satin coated cavies, which originate from the United States, are noted for the wonder-

Crème English crowned cavy.

Satin-haired English crowned cavy.

ful deep lustre of their coat. Their coats are short and need no more care and attention than normal-haired cavies.

Special remarks

New varieties of cavy have been created by cross-breeding rex, long-hair, and satin-coated cavies, such as satin-coated rex and satin-coated long-hairs.

These types are currently bred on a small scale but are not yet recognised in every country. This is also true of the English and American crowned varieties with satin coats.

English crowned cavies

Properties

English crowned cavies are extremely popular in a number of countries. The "crown" of hair creates an unusual appearance.
The difficulty in breeding this type is in getting the crown in the correct place – in the centre of the forehead – and as large as possible. The crown is often rather small or has several centres from which it fans out, which is a fault.
English crowned cavies are comparable in terms of their care with smooth-haired and satin-coated cavies, so that they need a minimum of attention. There are no differences in character between this type and other cavies. The crown is genetically dominant so

that mating between a smooth-haired and an English crowned cavy produces offspring with crowns.

External characteristics

English crowned cavies have the same body conformation as smooth-haired cavies.

Coat

The English crowned cavy is smooth-haired with little undercoat. The hairs are about 30mm (1¼in) long and they lay smoothly against the body.

The feature of this type is the crown of hair in the centre of the forehead. This should be as large as possible, with all the hairs fanning outwards from one centre. The colour of the crown is the same as the rest of the body, without any second colour.

Self black English crowned cavy.

Golden agouti English crowned cavy.

Colours

English crowned cavies are bred in every manner of colour but the most usual colours are the agouti colours and self-coloured black, red, white, and cream.

American crowned cavy

Properties

American crowned cavies are popular in a number of countries. Breeding this type is not easy because a show specimen must have a perfectly formed and correctly sited crown in the centre of its forehead and be pure white, without any other colour.
The white must be restricted solely to the crown and the body colour must not intrude

Red American crowned cavy.

crowned cavy will produce offspring with the crown.

External characteristics

American crowned cavies are no different in terms of body conformation than other cavies. Their bodies are also stocky, well covered with flesh, and muscular with well-rounded front and rump. The legs are short, muscular, and straight.

Coat

Golden American crowned cavy.

The difference between the American crowned and English crowned cavy is that the American crowned has a white crown that contrasts with the rest of its body colour.

Judges look closely at shows to ensure that the entire crown is white. The coat has similar properties to the smooth-haired and English crowned cavies, in other words, it has little undercoat, lays smoothly, is shiny, and the hairs are about 30mm (1¼in) long.

Colours

into the crown either. American crowned cavies have the same nature as other cavies and are short-haired, which is easy to care for.

The crown is genetically dominant so that mating a smooth-hair with an American

American crowned cavies are bred in a wide range of colours, including self-coloured red, black, buff, and golden – all of course with a white crown. The most common colour for this type is red.

Red Abyssinian cavy.

Brindle Abyssinian cavy.

Tri-coloured Abyssinian cavy.

Mixed roan Abyssinian cavy.

Abyssinian cavy

Properties

The Abyssinian breed of cavy originates from Britain. These are the most popular type of cavy after the smooth-haired types, especially for keeping as pets, but they are regularly to be seen at shows. Breeding good show specimens of Abyssinians is no easy task because of the value placed by judges on the right quantity, position, and form of the rosettes in the coat.

Two animals with perfectly formed rosettes cannot be guaranteed to produce offspring of the same quality, so that luck plays a role alongside knowledge in their breeding. Their are other specific characteristics with this breed too. Show specimens must have ears that hang down attractively, although these are initially erect with the young animals.

Only adults develop this desired form of ear. These cavies are identical in their care requirements to other cavies, because although the hair of the coat is longer, it does not tangle.

The wire-haired gene that produces the characteristic coat of Abyssinians is dominant so that if an Abyssinian is cross-bred with another cavy, the offspring will always inherit the Abyssinian coat properties, although the rosettes from such pairings are usually less well defined.

External appearance

These cavies have the same body conformation as other cavies but the siting of rosettes, particularly on the hindquarters, makes the shoulders less prominent in comparison with the hindquarters.

Rosettes on the nose.

Abyssinian cavy.

Coat

The hair of an Abyssinian cavy's coat is longer than that of smooth-haired examples, measuring about 35mm ($1^3/_8$in), feeling quite hard to the touch. The rosettes are found over the entire body. An ideal show specimen has four rosettes on each side of its body, placed symmetrically. This perfect specimen also has four more rosettes sited on its rump.

The shape and size of the rosettes is also important on the show bench. Judges prefer large round rosettes with ample hair that all radiate outwards from the centre of the crown. Judges like to see a rosette on the nose as well and a ridge of hair standing up along the back as a result of the rosettes on each side.

The bristly nature of the hairs, together with its length means that the hairs stand up. It can take up to one and a half years before an Abyssinian cavy is seen at its best. For this reason, younger specimens are rarely seen at shows.

Colours

Abyssinian cavies can be bred in all manner of colours. The majority, however, have so far been mainly bred in self red, black, and

Self black rex cavy

Tri-coloured rex cavy

Marked rex cavies

Pink-eyed white rex cavy

white. Tri-colour, marked, brindle, and tortoiseshell Abyssinians all exist though and agouti and roan specimens can be found.

Rex-coated cavies

Properties

Rex-coated cavies are fairly new arrivals on the scene but they have already found an eager band of enthusiasts for them. These cavies have also already won their way into homes as pets.

The mutation that causes the curly hair with rex cavies is recessively inherited so that a cross between a rex and a smooth-coated cavy generally will produce smooth-haired offspring. Only when two animals that both carry the recessive gene will a proportion of their offspring become true bred rex. Rex mated with rex produces only rex offspring.

Rex coated cavies are known as a Teddy in the United States.

External characteristics

Rex cavies have the same body conformation as smooth-haired cavies.

Coat

An attractive rex-coated cavy has very coarse but curly hair that does not lay

Young tri-colour cavy

Yellow and silver agouti marked rex cavies

smooth, instead it stands erect from the body. Judges do not like to see any smooth-haired parts of the coat on the show bench. The hair should be dense and it is fairly short and springy to the touch.
The hair on the head is usually somewhat shorter than the rest of the body.

Colours

Rex cavies are bred in a wide range of colours but not all of them are recognised for showing.

Tri-colour rex cavies are be-coming very popular but self-colours such as black, white, and red are kept by many fanciers.

Special remarks

The genetic factors for rex have been combined with satin in some countries and in the United States, to produce rex cavies with exceptionally lustrous coats.

Peruvian (long-haired) cavies

Properties

Peruvian (long-haired) cavies are popular pets.
To keep the coat in good condition demands a great deal of care and attention on the part of the handler and for this reason, it is best not to keep Peruvians on sawdust, using hay instead. The coat has to be checked every day for straw and such like and groomed with brush and comb to prevent tangles.

Most people who show Peruvians wind the hair in tissue paper to prevent the hairs splitting but also because severe tangles are unpleasant for the creature.

If you keep Peruvian cavies, it is best to trim the hair shorter from time to time.
This keeps the coat in good condition without so much work. Show specimens are also

Black marked Peruvian cavy

Exceptionally fine Peruvian cavy

Marked Peruvian cavy

Red marked Sheltie cavy with satin coat

Tortoiseshell Sheltie cavy

Tri-colour Sheltie cavy

External characteristics

Peruvian long-haired cavies should have the same body conformation as other cavies. The form of the body is not as apparent with smooth-haired cavies though because of the long hair.

Coat

Peruvian long-haired cavies have shiny coats that feel soft to the touch. The only short hair is that at the front of the nose. Because of the rosette on their heads, the long hair falls forward to cover the nose when the animal gets older and the hair becomes much longer. There is also a "train" of hair flowing behind them from the rump. There is no standard for maximum hair length.

Colours

Peruvian long-haired cavies are bred in diverse colours, including self red, black, and white, plus tri-colour, with markings, and tortoiseshell.

Special remarks

Peruvians are at their best when young. When they are older, the long hair is often stiffer and has less shine. There are now also satin-coated Peruvians which have a won-

clipped. The long-hair gene is recessive, so that long-haired offspring can be born following a pairing between a long-haired cavy and a normal one but also two cavies that do not have long hair can produce long-haired offspring if they both carry the gene.

Two long-haired cavies also produce long-haired offspring.

derful sheen on their coat throughout their lives due to the satin influence.

Sheltie (long-haired) cavies

Properties

Shelties look much like Peruvian long-haired cavies but there is one important difference – there is no rosette on the head. This allows the hair to grow backwards instead of falling forward across the face. In common with Peruvians, Shelties need a great deal of grooming to keep their coats in good condition, so that they are not suitable pets for people with little time or inclination to brush and comb their cavy every day, checking through the coat for pieces of straw or any tangles.

Judges look for coats that are as long as they can possibly be but it is not easy to achieve

Two Shelties

Marked red Sheltie

An exceptionally fine marked red Texel cavy.

this ideal. The hair can be prevented from tangling and splitting by rolling it in strips of tissue paper. In common with all other cavies, Shelties are good natured animals.

External characteristics

The Shelties should have the same body conformation as all other cavies.

Coat

The Sheltie is a long-haired cavy which differs from the long-haired Peruvian by the absence of a rosette on the front of the head, so that the head does not fall forwards over the face. The Shelties does not have any rosettes on its rear either.

The hair falls either side of the back but there is hardly a question of a parting. The Sheltie has a beard on its cheeks that flows into the hair on the body. Judges like to see a well-developed "train" of hair flowing behind the animal.

Colours

Shelties are bred in diverse colours. The most popular are tri-colours (combinations of black, red, and white), black and red marked cavies, tortoiseshell (black and red), and self colours such as red, white, and black.

Special remarks

The Shelties looks its best under the age of two years. After this age, the coat loses some of its sheen and it starts to feel less soft to the touch. Shelties are now also being bred with satin coats which retain their lustre throughout the cavy's life.

Texel cavy

Properties

Texel cavies are not yet as widely adopted as Peruvians but their curly hair has won an audience for itself fairly quickly. The care of the Texel's coat is time consuming, in common with other long-haired breeds, because it is susceptible to tangles, but the coat also curls, which brings further difficulties for grooming.

Before a Texel is shown, its coat will require a thorough grooming with brush and comb to ensure it is entirely free of tangles. This

Tri-colour Texel cavy

can cause the hair to lose its curl so it is best to lightly spray water on the hair with a plant sprayer and to then curl the hair with the fingers. It is entirely out of the question to keep these animals on a bed of sawdust because it becomes caught up in the hair. It is better to use hay or straw.

External appearance

Texel cavies have the same bodily conformation as all other cavies, with the exception of their hair.

Coat

Texel cavies were first bred in Britain. They are really long-haired rex cavies. Their coat consists of long, wavy hair that is soft to the touch. The hair should shine and be dense, with no bald patches or areas in which the hair is less dense.
The hair on the head should grow towards the rear across the back. The hair on the

Tortoiseshell Texel cavy

nose is shorter than on the rest of the body. The Texel has a parting in the middle of its back.

The hair on the belly forms small curls. Judges look for a well-developed "train" of hair flowing out behind the Texel. The hair should be about 120mm (4¾in).

Colours

Texels are bred in various colours. The most readily available up to now have been self red and white. and red marked specimens but there are also Texels with colourpoint markings, with agouti colours, and tricolour but these are not all recognised in every country.

Special remarks

Texel long-haired cavies are now also bred with satin coats but these are not yet recognised and are extremely rare.

Merino cavies

Properties

Merino cavies need the same kind of care and grooming as Texels. This makes them less suitable to keep as pets for people who have too little time or are not prepared to devote the time needed to keep their coat in

Merino cavy

Merino cavy

good condition. Like Texels, it is really not possible to keep Merinos on a bed of sawdust because it gets caught in the coat and leads to tangles. Straw is a better bedding material.

External characteristics

Merino cavies should have the same bodily conformation as all other breeds.

Coat

These long-haired cavies have a curly coat with two rosettes, combining the features of the Peruvian with the curls of the Texel.

Colours

Merinos are bred in a number of colours, including self red, black, cream, and white, together with marked cavies and tricolours.

Special remarks

Merinos are now also being bred with the lustre of satin coats. Merino cavies are not well known with the public and few can be seen at shows because in many countries they are not yet officially recognised.

Coronet cavies

Properties

Coronet cavies require similar grooming and care as Texel cavies and are less suitable for keeping by people who have little time for or interest in daily grooming with brush and comb. Sawdust is not suitable for use on the floor of their cage because it becomes enmeshed in their coat. Straw is more suitable bedding material for this breed.

External appearance

Coronet cavies should be stockily built like all other cavies, with a muscular frame and well rounded front and hindquarters.

Coat

The coronet cavia is a Sheltie with a rosette on its forehead. The hair on the head is short.

Colours

Coronet cavies are bred in a number of colours, including self red, white, black, and cream, plus marked coronets, and also tricolour.

Alpaca cavy

Special remarks

Coronet cavies are now also being bred with satin coats to produce coronets with a deep lustre to their coats. Both varieties are not yet fully recognised in a majority of countries and they are both very rare.

Alpaca cavies

Properties

Alpaca cavies require similar grooming and care as Texel cavies and are less suitable for keeping by people who have little time for or interest in daily grooming with brush and comb.
Sawdust is not suitable for use on the floor of their cage because it becomes enmeshed in their coat. Straw is more suitable bedding material for this breed.

External appearance

Alpaca cavies should have the same bodily conformation as other cavies.

Coat

The coat has long, curly hair, rather like the Texel cavy. The Alpaca cavy is differentiated from the Texel by the rosette on its forehead.

Colours

Alpaca cavies are bred in diverse colours, including self red, cream, black, and white.

Special remarks

Alpaca cavies are now also being bred with satin coats to produce coronets with a deep lustre to their coats. This variety is not yet fully recognised in a majority of countries and is very rare.

10. Rabbit breeds

Large rabbits

Flemish Giant (Flemish)

Origins

The Flemish Giant, with its body length of 800mm (31½in) and weight of about 7kg (15½lb) or more is the largest sized breed of rabbit in the world. The breed was being raised in Belgium to a breed standard and fairly strict pedigree as early as the nineteenth century. It is generally considered that the Flemish Giant was derived from another large breed of rabbit which has died out: the Patagonian rabbit. The Patagonian rabbit was once fairly common in Belgium and also in France. Various sources suggest that the Flemish Giant was already being pure bred in the sixteenth century in the area surrounding Ghent under the name of the Ghent Giant. There were even various clubs involved in the breeding of these rabbits. Unfortunately so little has been preserved in writing about the breed from this period, so that we know little about its development until the nineteenth century.

Left: young rabbit grey Flemish Giant

Rabbit grey Flemish Giant doe

Three weeks old rabbit grey Flemish Giants

The name *Vlaamse reus* (Flemish Giant) first appears in writing in the middle of the nineteenth century. These rabbits were bred on a massive scale but the breeding remained centred around Ghent, in the Flemish speaking part of Belgium. It appears that by 1880 there was interest in this breed in other countries, judging by the first export to Germany registered in that year.

Originally, the Flemish Giant was solely bred in its natural hare-like colour and steel grey. The white examples did not appear until later, and because they are of somewhat smaller build, they were treated as a separate breed for showing purposes.

The white Flemish rabbits are still often judged separately from the coloured Flemish Giants, although the weights are now fairly standard in most countries.

Further colours were created by cross-breeding with other breeds such as Giant Papillons and Beverens. Subsequently, numerous coat colours have become known

with enthusiasts for Flemish Giants throughout the world.

Flemish Giants can weigh more than 7kg (15½lb)

Properties

Although Flemish Giants are tremendously popular with breeders and can usually be seen in large numbers at shows throughout the world, they are not much sought after as children's pets. The disadvantage of the Flemish Giant is that they cost more in upkeep than other breeds. They need a spacious run and hutch, eat relatively a lot of food, and are not easy for children to manage because of their weight.
The majority of Flemish Giants are characterised by their good natured, reliable, and calm behaviour.

External characteristic

The Flemish Giant is the largest breed of rabbit in the world. The general minimum weight for the breed is 5.5–6kg (12lb 2oz–13lb 4oz) but judges prefer to see much larger specimens than this on the show bench, provided the weight is not the result of excessive feeding or too fatty food. The average weight for Flemish Giants is 7–8kg (15½–17½lb). Some countries treat the white Flemish Giants as a separate breed

Young black Flemish Giant

Steel grey Flemish Giant

Fine white Flemish Giant

which permits a lighter rabbit than the coloured counterparts. Flemish Giants have long bodies and a wide back.

The length of the back is at least 650mm (25½in). The legs are of average length, strong and muscular, and straight. The back should be straight with a well-rounded and solid fleshed hindquarters. The head is powerful and wide with full cheeks. There is a clear difference between the head of a buck and that of a doe. The female's head is less imposing. The ears are strong, large, and thick, being held in a Vee-shape straight above the head. They measure about 180mm (7in) long.

Coat

Flemish Giants have a coat of normal-length hair with ample undercoat. The standard requires the coat to lay smooth with a glossy sheen.

Colours

Flemish Giants are bred in various colours such as agouti and self colours. Spotted and marked specimens are not approved in this breed standard.

AGOUTI

The natural wild agouti colouring is characterised by different colour bands on the hairs. The colour closest to the body is known as the undercolour. This is grey with many animals. The colour at the top of the hairs is known as ticking.

The ticking is usually black or blue. The colour between these two is the base colour. This is usually light brown or grey with slight differences in shading. The three different colours can be seen if one blows the hairs. Rabbits with agouti (or natural wild) coloured coats also have light coloured bellies, insides and backs of their legs, undersides to their tails, and light rings around the eyes. The tips of the ears are the darkest part of the coat. Agouti can be further divided into hare coloured (reddish base colour with black ticking and brown eyes), rabbit grey (light grey/brown base colour with black ticking and brown eyes), blue-grey (light brown base colour with blue-grey ticking and blue eyes), steel grey (light grey base coat with black ticking and brown eyes), and blue/dun (light grey base coat with blue ticking and blue eyes). Argente is yellow coat half-way between agouti and self colour. Rabbits with this colour in reality have an agouti colour without any ticking. They also have light coloured bellies and light rings around the eyes. Some argente Flemish Giants exhibit slight ticking but this is considered a fault.

Self-coloured

Self-coloured Flemish Giants have no ticking or other characteristic from the natural rabbit coat. The hair is entirely one

Rabbit grey Flemish buck

Argente Flemish Giant

Rabbit grey Flemish doe

Blue Flemish Giant

White British Giant

colour without the presence of hairs of any other colour. The belly and area around the eyes is of the same colour as the rest of the body, perhaps less shiny, but certainly not lighter.

The principal self colours for Flemish are blue with blue eyes, black with brown eyes, and albino (white with pink eyes).

Special remarks

Flemish Giants are normally referred to in Britain as Flemish and their weight is generally somewhat lower (the minimum weight in the breed standard is 4.9kg (11lb). Only dark steel grey rabbits with even ticking over the entire body are accepted in the United Kingdom.

British Giant

Origins

The British Giant is a breed that appeared in the 1940s in the United Kingdom, where only steel grey Flemish rabbits were recognised – and that is still the case today. Some breeders wished to introduce other colours, partly because they considered the strain of Flemish rabbits in Britain to be too small, which they felt might be associated with the colour.

They decided to import Flemish Giants of different colours from the United States in order to breed them as a separate breed. The first imports arrived at the end of the 1940s and the British Giant Club was founded at that time too.

This club sub-sequently ceased to be, which also spelled a down turn in the development of the breed but much later in 1981, a new club was founded to further the breeding of the British Giant. Because Flemish Giants are bred all over the world in a range of colours, there is little impetus to encourage the export of British giants. They are therefore virtually unknown outside the United Kingdom.

Properties

The British Giant is a friendly breed that is very popular in Britain with rabbit fanciers. The breed is less popular though as household pet for children because of its size.

External characteristics

British giants have broadly the same characteristics as the Flemish Giant but they weigh less: the minimum weight for does is 6.1kg (13½lb). The minimum weight for bucks is 5.6kg (12lb 5oz).

Coat

The British Giant has a coat of normal length hair with a dense undercoat.

Colours

The British Giant is recognised in black, blue, rabbit grey (light brown base colour with black ticking and brown eyes), and steel grey (light grey base colour with black ticking and brown eyes). White rabbits with pink or blue eyes are also recognised.

Blanc de Bouscat

Origins

Blanc de Bouscat is a French breed of rabbit, bred by Paul Dulon. It is said that the

breed resulted from cross-breeding Flemish Giants, Angora rabbits, and Argenté de Champagne. By solely breeding with the albino offspring from this breeding, a breed of pink-eyed white rabbits has been developed. The breed has been seen regularly at international shows since 1910 but it was not recognised in France until 1924.

The breed does not have a wide band of enthusiasts outside its native land and is extremely rare in the United Kingdom and unheard of in the United States.

Properties

Blanc de Bouscat is a generally calm breed of rabbits with even temperaments. The animals are almost exclusively kept by rabbit fanciers.

External characteristics

Blanc de Bouscat closely resembles the Flemish Giant but has a more elegant body. The body is extended, with a curved back. The head is well formed with a lightly curved nose profile.

The ears are proportionately large at 150–180mm (6–7in) and stand erect above the head in a Vee-shape. The minimum weight for this breed is 5kg (11lb) but most examples are slightly heavier than this.

British Giant and Polish: the largest and smallest of British breeds.

Havana brown Giant Papillon

English bred Giant Papillon

Young Giant Papillon

Blue Great Lorrainese

Black Giant Papillon doe

Coat

The breed standard sets a minimum length of 30mm (1¼in) for the hair.

Colours

Rabbits of this breed have pure white coats and always have pink eyes.

Giant Papillon (Great Lorrainese/American Checkered Giant)

Origins

The largest breed of rabbit after the Flemish Giant is the Great Lorrainese (known in Britain as Giant Papillon) which came into existence towards the end of the nineteenth century in the then German region of Lothringen (today's Lorraine in France) by cross-breeding Flemish Giants with large French lop-eared rabbits, and spotted rabbits that were kept in France at that time.

Giant Papillon buck

The moustache-like marking of a "Charlie".

A litter of Giant Papillon can include plain colours but also predominantly white rabbits.

The original intention of this breeding was not to produce an attractive rabbit but to develop a new rabbit for its meat with the length of the Flemish Giant, the solidity of French lop-ears, and the resistance of commercially bred rabbits for the meat trade.

The breeders obviously succeeded because the predecessors of today's Great Lorrainese were 6–7kg (13¼–15½lb). The characteristic markings of today's Great Lorrainese were then not in evidence: the majority of those rabbits were multi-coloured or natural wild coloured. Some of these animals were sent to the rest of Germany where they were further devel-oped and given the name *Deutsche Riesenschecke* (Great German Spotted). The breed was kept in most European countries from the 1920s onwards.

The first Great Lorrainese were imported into the United States at the beginning of the twentieth century and became known as American Checkered Giants, recognised in just black and blue. The British version of the Great Lorrainese is the Giant Papillon) which was not recognised in the United Kingdom until 1994. The breed may be bred in Britain in any colour.

Properties

The Giant Papillon (Great Lorrainese / American Checkered Giant) is a very calm rabbit which, with its distinctive markings, gets a great deal of admiration at rabbit

shows. This breed is marked and therefore judges pay special attention to the quality of the markings at shows. The smaller and more manageable English/English Spot/Papillon is more popular as a children's pet.

External characteristics

Great Lorrainese (Giant Papillon/American Checkered Giant) are large rabbits that look a great deal like Flemish Giant in terms of their build, yet more elegant.In common with the Flemish rabbit, this breed also has a long body with a straight back. Judges pay close attention to the length of their ears: these should be a minimum of 160mm (6¼in) and be held firmly erect. The minimum weight for Giant Papillon is 5kg (11lb) but there is no maximum weight.

Coat

The coat of this breed should be short, dense,

with an ample undercoat. The hairs should not stand up but lay smooth with an attractive sheen.

Markings

The markings on an otherwise white coat are characteristic for this breed. The markings on the head should include the dark "butterfly" on the nose (from which the Giant Papillon name is derived), rings around the eyes, a "thorn" marking, and markings on the cheeks. The "butterfly" should be clearly defined and stretch to both corners of the mouth. The eye rings should be of even size. The "thorn" is a marking right in the middle of the top of the nose. All these markings are essential to comply with the standard but local variations can apply. The Giant Papillon also has a dorsal stripe on its body, which runs from behind the ears to the tail. Judges look for an even thickness and sharp definition of this marking. Finally, these rabbits also have several markings on either side of their body, which are ideally round spots that should not merge with each other. Any white hairs within the dark markings is considered a fault.

Colours

These breeds come in a variety of second colour for the markings. The most predominant colour was always and still is black but rabbit grey (light brown/grey base colour with black ticking and brown eyes), blue/grey (light brown/grey base colour with blue/grey ticking and blue eyes), steel grey (light grey base colour with black ticking and brown eyes), and blue/dun (light grey base colour with blue ticking and blue eyes) are all recognised for the Great Lorrainese. This breed is also bred with havana brown (with brown eyes), blue (with blue eyes), amber (with brown eyes), dun (with blue/grey eyes), and tri-colour.

Special remarks

The breeding of perfectly marked specimens of these breeds is not a simple matter, since the mating of two perfectly marked specimens does not guarantee successful offspring. Each litter will virtually always contain some plain coloured rabbits and some that are predominantly white but without the correct markings as well as perhaps rabbits that are correctly marked. Some suggest that broken butterfly markings acquired the name Charlie from Charlie Chaplin's moustache. Whether true or not, the term has wider use for any animal in a litter that deviates from the desired colour or marking.

Blauwe van Sint Niklaas (Bleu de St Nicholas)

Origins

The "St Nicholas blue" is an old Belgian breed that originates from St Niklaas, where it was first bred towards the end of the nine-

Young Blauwe van St Niklaas

Blauwe van St Niklaas

teenth century. It is not known which rabbit breeds (perhaps even wild rabbits) were used to produce this breed but it is quite likely that blue Flemish Giants were involved but blue Beverens were almost certainly involved. The breed is virtually unknown outside Belgium and France.

Properties

The "St Nicholas blue" is generally a calm, good-natured, and friendly rabbit. The breed is mainly kept by rabbit fanciers because the size and weight make them less suitable as household pets.

External characteristics

The *Blauwe van Sint Niklaas* has an extended body with a fairly straight back and strong, straight legs. The head is fairly long with ears that have an average length of about 130mm (5⅛in). Rabbits from this breed weigh 4.5–6kg (9¼13¼lb).

Coat

The coat is medium length with a dense undercoat and it lays smoothly against the rabbit's body.

Colour

The colour of the coat with this breed is always blue. The blue must be of a similar tone and depth of colour tight to the root. The eyes are blue/grey.

French Lops have a calm temperament.

Young steel-grey marked French Lop

Young rabbit-grey French Lop

Lop-eared breeds

French Lop

Origins

The French lop is a long-established breed in France where it was bred for the table from the middle of the nineteenth century. In those days, standards of hygiene were less well developed and rabbits frequently suffered from canker in the ears. The French breeders realised that the English Lop, which has been present in France from the beginning of the nineteenth century, rarely suffered from this annoying problem. The breeders decided to cross breed English Lops with wild rabbits and the predecessors of the Flemish Giant in order to create a good meat producing breed that was strong, with the desired lop ears.

The rabbits that were produced from this programme were lighter than the enormous French Lops that we see today on show benches. They probably were without the white patches, indeed they were predominantly rabbit grey. The natural colour of wild rabbits was in those days by far the most in demand for rabbits for the table because they were regarded as providing better meat than other colours. The French Lop with markings was created by German breeders, who regarded the French Lop highly. These breeders crossed the then French Lop with non-pedigree spotted rabbits. The French Lop became a very popular utility breed in Germany, meaning it was principally bred for its wool-like pelt and meat. By the end of the nineteenth and beginning of the twentieth century, virtually every country in Europe except Great

Britain was busy with rearing this breed. The breed did not arrive in Britain until 1938 when they were also exhibited for the first time in the United Kingdom at Crystal Palace in London. Interest among rabbit fanciers in those early years was not very high and the breed was almost exclusively reared for the table but the French Lop was also used to improve the English Lop. The breeder Meg Brown started to show these

Albino French Lop

Blue French Lop

rabbits extensively from 1965 and her input had an enormous influence in the recognition of the French Lop in the world of rabbit shows. Subsequently, the breed has become known at shows throughout the world.

Properties

French Lops are very calm, good natured creatures with an even temperament. They were originally bred for the table and the young rabbits develop extremely quickly and are soon fully grown.

This remarkable breed has the nicknames of "bulldozer" and "bulldog" among its enthusiasts and it is still bred for the table but has become increasingly popular on show benches as well where it is much admired.

The breed is rather too large to find a place as a household pet.

External characteristics

The French Lop, which weighs in at between 4.5 and 6kg (9¼–13¼lb), belongs with the large breeds. These rabbits have a sturdy, even stocky and muscular body with a broad chest.

The hindquarters are broad and well rounded. The legs of this breed are fairly sturdy and short. The French Lop has a generously proportioned, broad, and solid head, which is particularly wide between the eyes.
A curved profile to the top of the nose is an important feature of the breed. The head is almost always larger with bucks then does. Judges not only pay attention to the overall proportions of the body, they exam-ine the ears for shape, length, and overall size.

The ears should hang down beside the head without any folds. The inside of the ear should not be exposed but should lay at the side of the head. The thick and fleshy ears are nicely rounded at their tips. Where they

Blue/grey French Lop

Rabbit grey French Lop

A "Roman" profile for the nose is characteristic of the French Lop.

Amber French Lop

are attached to the head, lumps are clearly visible. These "crowns" are a desirable feature for showing. The average length of the ears from end to end is about 420mm (16½in).

Coat

French Lops have a dense shiny coat with a substantial undercoat. The hairs should feel soft to the touch and somewhat longer than normal-haired breeds.

Colours

French Lops are bred in a variety of colours, including rabbit grey (light grey/brown base coat with black ticking and brown eyes), blue/grey (light grey/brown base coat with blue/grey ticking and blue eyes), steel grey (light grey base coat with black ticking and

brown eyes), chinchilla (ivory base coat with black ticking and brown eyes), and blue/dun (light grey base coat with blue ticking and blue eyes). In addition to these colours, there are also self black and blue, amber, and creamy dun. French Lops with markings are found in manner of combinations but most are black or rabbit grey. Specimens with a white chest, white legs, and a white belly are highly prized at shows, although judges tend to be more interested in the symmetry of any white markings than their extent. There are also white specimens (albino with pink eyes and white with blue eyes).

Special remarks

The ears of young French Lop rabbits are erect at first. The ears start to hang down when they are four weeks old and this does not happen uniformly: normally one ear drops first.

Meissener Lop

Meissener Lop

Origins

The Meissener Lop is a German breed of rabbit that was developed by a breeder from Meissen. Towards the end of the nineteenth century Mr. R. Reck crossed lop-eared rabbits with, it is assumed, Argenté de Champagne, and Silvers, to create a breed of lops that bear the silver gene. Since Reck kept the "formula" for the Meissener Lop to himself, it is impossible to be certain of the blood-lines utilised.

Meissener Lop

The first example from this breeding to be shown was a silver-black rabbit that was shown as a "Meissner Widder" in 1906. The black colour was the only colour recognised at first but other colours were added subsequently by crossing with other breeds.

This Meissener Lop is moulting heavily.

Properties

Meissener Lops are generally calm and affectionate natured rabbits. The breed is virtually unknown to the public but has a small loyal group of enthusiasts among breeders. The breed is virtually restricted to Germany and a few countries surrounding it.

External characteristics

Meissener Lops have bodies that are slightly extended, with short and sturdy muscular legs. Their bodies are broad and well fleshed. The neck is very short and above all muscular and thick.

The back is gently arched with broad, generous hindquarters that are well rounded. The head is substantial and wide, with full cheeks.
The distance between the eyes is wide and the nose is curved. Bucks usually have a more pronounced head than does.

The substantial ears should hang down straight beside the head by they should not in any way be limp, nor have any creases or folds in them.

There are distinctive bumps where the ears are attached to the head. These "crowns" are a required breed characteristics. The end of each ear should be nicely rounded off.

do not apply to this breed. The challenge is to produce rabbits that have the silver evenly distributed across the body but generally, rabbits of this breed have little silver in their head, legs, and ears. The eye colours are derived from the coat colour.

Special remarks

The young rabbits of this breed initially have plain coats. The silvering in the coat does not start to develop until they are five to six weeks old.

English Lop

Origins

The English Lop is the oldest of the lop-eared breeds. Virtually every other lop-eared breed of rabbit has been derived from the English Lop. Rabbits resembling the English Lop were first known in Britain in the eighteenth century.

Correspondence and books of the time mentioned rabbits with extremely long and broad ears. These rabbits, almost all with white or marked coats, were bred more or less without outside stock, under the name Lops.
The breed is extremely popular in Britain where enthusiasts constantly strive to increase the length of their ears. In addi-

The ears should measure 380–420mm (15–16½in) from skull to tip. The weight of the Meissener Lop is 3.5–5.5kg (7lb 12oz–12lb 2oz).

Coat

Great value is placed at shows on the quality of the pelt of the Meissener Lop. This should be dense and soft with a lustrous sheen. The hairs of the coat do not stand up but should lay smoothly against the body and should be somewhat longer than that of short-haired breeds.

Colours

This breed is bred in the same colours as the small Silver: rabbit grey, black, argente, blue, and brown. The extent of silver influence in the coat (light, medium, or dark)

English Lop in the popular fawn colour

Black English Lop

tion to continually selecting stock with the longest ears for breeding, the rabbits were also kept in heated hutches because it was believed this helped the development of the ears. This method is no longer used because rabbits that have for generations been kept in such accommodation prove to be weaker and more sickly than rabbits that have had the opportunity to develop resistance. The English Lop has also found its way onto the international show scene.

Properties

English Lops have loyal enthusiasts who have remained with the breed for many years even though interest is much lower in the breed in some countries than others. It is not easy to breed specimens fit for showing because of the rigorous breed standards that apply. It seems to be best to produce the offspring in the summer months because the warm weather seems to encourage rapid growth of the ears. The breed is not widely kept as a pet because of the additional care demanded by the unusual ears. Those who keep English Lops also have to regularly clip the claws on the front feet to prevent the rabbits from hurting their ears with their claws.

External characteristics

There are two characteristics of the English Lop that set it apart from other breeds. The length and breadth of its ears are not developed to such an extreme in any other breed and the way the line of the back increases from front to back with this slender rabbit catches the eye too.

The English Lop is slender yet solidly built with legs that are shorter at the front than the back. Both pairs of legs should be straight. The head is carried low set and should be broad, especially with bucks. These rabbits have a fairly long head with arched nose profile. The insides of the ears face forwards, with the broadest part in the centre of the ear, terminating with a rounded tip. Judges check the length of the ears at shows but also pay attention to whether they are damaged in any way or bear scars from earlier injuries.

The ears are normally measure from ear tip to ear tip, including the skull. The average measurement lies in the range 580–700mm (22⅞–27½in). The minimum width of the ear, measured at its widest point (in the middle) should be 120mm (4¼in) but judges prefer to see much thicker ears. No maximum width is given in the breed standard. Rabbits attain their adult size of ear when they are about five months old. English Lops weigh 3.5–5kg (7lb 12oz–11lb).

Coat

English Lops have a fairly short coat with adequate undercoat. The hairs feel soft to the touch, shine, and lay smoothly against the body.

Colours

English Lops are bred in various colours. By far the most usual colour is fawn but they are

Chinchilla German Lop

Rabbit grey German Lop

also found in rabbit grey (light brown/grey base coat with black ticking and brown eyes), steel grey (light grey base coat with black ticking and brown eyes), and black with brown eyes. English Lops with colour markings are not uncommon.

These have a white chest, belly, and hind legs, with one or more patches on the head. Albino English Lops occur occasionally.

Special remarks

A black English Lop doe named "Sweet Majestic Star" was entered in The Guinness Book of Records in 1994 for the longest rabbit ears of all time. From ear tip to ear tip, including the skull, she measured 724mm (28½in).

German Lop

Origins

The German Lop is a recently arrived breed from Germany. The breeders used Holland Lops and French Lops among others with the intention of creating a breed like the French Lop with a size and weight mid-way between these two breeds. The overall impression of the German Lop is far more of comparison with the French Lop than the Holland Lop.

The resulting breed was officially recognised in Germany in 1970, where it was given the

name Deutsche Klein Widder. The breed quickly become popular in Germany.

The breed was imported into The Netherlands in 1972 and officially recognised there in 1976, and has subsequently become a popular breed to be seen at virtually every show.

The spread of the breed beyond these countries is limited, although there is a loyal band of enthusiasts for the breed in the United Kingdom, where they were first introduced during the 1980s by the Dutch breeder Ms E. van Vliet, who took them with her, when she moved to England. Together with her English colleague David Cannon, who took over some of her rabbits, they were instrumental in getting the breed recognised in Britain, obtaining the coveted recognition certificate in January 1990.

During the early years of the German Lop, marked specimens were quite rare. The marked strain has been imported to other countries from Dutch stock.

Exceptionally finely marked German Lop in rabbit grey.

External characteristics

The German Lop closely resembles the French Lop but they are smaller. The weight of a German Lop is 2.5–3.5kg (5lb 8oz–7lb 12oz), while the French Lop weighs 5kg (11lb) or more. The breed is characterised by its stocky, compact, and sturdy body. The hindquarters is firm and well rounded.

The neck is very short and the legs are short, muscular and powerful. German Lops have a broad, short head, with noticeable cheeks, and a curved nose. The ears measure 300mm (11¾in) from tip to tip, including the skull. The ears should hang straight down, without any folds. This breed has highly developed "crowns" where the ears leave the skull.

Coat

The coat of a German Lop is slightly longer than most rabbits, in common with the French Lop, and its is dense with an ample undercoat.

Properties

German Lops are calm and good natured. They are principally bred by rabbit fanciers for showing but their gentle nature and unusual appearance make them ideal family pets for both children and adults.

Colours

German Lops are bred in many different colours including: rabbit grey (light grey/brown base coat with black ticking and brown eyes), blue/grey (light brown/grey base coat with blue/grey ticking and blue eyes), steel grey (light grey base coat with black ticking and brown eyes, blue/dun

Holland Lop

Origins

The Holland Lop is a Dutch breed that was almost simultaneously, but independently of each other, bred by a well-known Dutch rabbit judge and a Dutch breeder.

The breeder, a Mr. A de Cock is credited in publications with the most important role in the development of the breed. The plan to create a miniature version of the French Lop was begun in 1952. He crossed a French Lop doe with a blue/sable Netherlands Dwarf.

The resulting offspring were a disappointment in that not one had lop ears and a further crossing between one of the progeny and a French Lop was equally unsuccessful. By the introduction of English Lops, Mr de Cock was more successful, although many of

Blue dun Holland Lop

(light grey base coat with blue ticking and blue eyes), and chinchilla (ivory base coat with black ticking and brown eyes).

There are also self coloured German Lops in black, blue, and white with both pink eyes (albino) and blue eyes, plus two colour marked specimens in amber and dun.

Special remarks

Because of their weight, the ears of this breed hang down soon after they have left the litter.

It is almost unheard of for the ears to remain erect with this breed, as can happen with the Holland Lop.

Exceptionally fine amber coloured Holland Lop

Black Holland Lop

Blue Holland Lop

Rabbit grey Holland Lop

Amber coloured Holland Lop

the rabbits born had one ear up and one ear down, and some had both ears erect. It took him twelve years of careful selection and significant effort to achieve the smaller lop-eared rabbit he sought.

The new breed was shown by him in 1964 for the first time and was officially recognised in that same year by the Dutch Rabbit Breeders' Association. The breeder Schrey

started his quest for a miniature French Lop about 1962. He developed his breed using Papillons, Netherlands Dwarfs, and French Lops. In 1966, he exhibited the resulting rabbits for the first time. Initially, the rabbits were mainly amber coloured, rabbit grey, and blue dun.

The other colours were developed subsequently. The breed found its way into the

Black and white Holland Lop

Albino Holland Lop

Dun coloured Holland Lop

Two young Holland Lops

Blue/grey Holland Lop

United Kingdom in 1968 and was officially recognised in 1976. The first Holland Lops were exported to the United States in 1969, where the albino and sable types were in particular soon in great demand.

Properties

Holland Lops are friendly, good-natured little rabbits that fall somewhere in the middle of the spectrum for liveliness. These rabbits can be seen on show benches all over the world, in virtually ever country where rabbits are kept and bred for pleasure. In many countries, they are one of the top ten most popular breeds.

The convenient size, friendly and lively nature, and amusing appearance make this rabbit an ideal household pet for children. Breeding specimens that comply with the breed standard is not at all easy. Many examples are still born that are too large or that grow up to be too big, so that they can no longer be considered a miniature rabbit.

A littler of two-coloured Holland Lops

Holland Lops are most widely bred in amber-coloured "Madagascar".

The position of the ears also causes considerable problems. The ears of the young rabbits do not droop immediately, this can take four to sometimes even twelve weeks to happen.

If the ears are to short, or wrongly positioned, they can even remain erect, or stick out sideways. Sometimes one ear sticks up while the other hangs down.

It is usually a good sign if a young rabbit's ears droop at a very young age because such examples normally develop into excellent showing specimens.

External characteristics

Holland Lops should look like a miniature French Lop. The average weight for these rabbits is about 1.5kg (3¼lb) but there are also lighter and heavier examples.

These rabbits have a short, stocky and sturdy build. Their necks are short and so too are their strong, thick legs. In common with the French Lop, this breed should have a substantial, broad but short head, with a curved nose profile.

Where the ears exit the skull, there should be bumps or "crowns," which result from the bending of the ears. Measured from tip to tip, including the skull, the ears should be 210–260mm (8¼–10¼).

The ears should hang down straight with the hairy side outwards, without any twisting. The ears should be sturdy and have fine rounded points.

Coat

Holland Lops have a densely-formed coat that is somewhat longer than Netherlands Dwarfs. The undercoat is very woolly and the hairs feel soft to the touch.

Colours

Holland Lops are bred in a wide assortment of colours, including rabbit grey (light brown/grey base coat with black ticking and brown eyes), blue/grey (light grey/brown base coat with blue/grey ticking and blue eyes), steel grey (light grey base coat with black ticking and brown eyes), and blue dun (light grey base colour with blue ticking and blue eyes.

Holland Lops are also bred in self black, blue, and both white with blue eyes and

albino (white with pink eyes). There are also examples with second colour markings in dun, various sable varieties, silver fox, and Sallander rabbit colouring.

Amber or "Magagascar" is by far the most popular colour. Finally, there are also Holland Lops with markings. The most eagerly sought after for showing are those with the chest, legs, and belly of white and with no or virtually no white hairs in the coloured parts of the coat.

Special remarks

There are two versions of this type of rabbit in the United Kingdom. The Holland Lops have a weight of 1.8–2.5kg (3lb 15oz–5½lb) and must not exceed this upper weight.

Mini Lops (which combine Holland Lops and other dwarf lop-eared breeds) have a weight range of 1.5–1.7kg (3lb 5oz–2lb 7oz).

Cashmere Lop

Origins

The Cashmere Lop is of British origins, having been developed from Holland Lops and English Angora rabbits.

The first long-haired Holland Lops were seen in Britain in at the beginning of the 1980s and within a few years they were recognised as a separate breed. The breed is

A Giant Chinchilla rabbit's coat is longer than normal.

virtually unknown outside Britain but because for many, their combination of small size, lop ears, and medium length, soft hair is irresistible, it can only be a matter of time.

External characteristics

The Cashmere Lop has the same external characteristics as the Holland Lop, except for its longer hair, which feels extremely soft to the touch.

These rabbits weigh about 1.5kg (3lb 5oz). Cashmere Lops are bred in a wide assortment of colours such as self black, white, and amber, plus chinchilla (white base coat with black ticking and brown eyes), and various shades of sable.

Medium-sized breeds

Giant Chinchilla

Origins

Rabbits known as Chinchillas already existed in France at the end of the nineteenth century. These rabbits weighed about 3kg (6lb 10oz) and were principally bred for their pelts. Work started on developing a Giant Chinchilla in both England and Germany at the same time, where breeders each had their own ideas about how to create a larger version of the French Chinchilla rabbit. In Germany, Offenbach, Grüny, and Geyer from Ilmenau achieved their first goal by crossing albino rabbits with selected agouti rabbits.

In England, Chris Wren spent the first part of the 1920s crossing standard-sized Chinchilla rabbits from France with selected Flemish Giants. Although today Chinchilla is taken to mean a rabbit with an ivory base coat with black ticking, Chinchilla rabbits also used to be other colours.

Properties

The Giant Chinchilla is an exceptionally calm rabbit with an even temper. These creatures produce fairly large litters and grow extremely rapidly. The breed was once sought after for the table and for its fur. Giant Chinchillas are now principally kept by rabbit fanciers who wish to improve the breed and to show them.

External characteristics

The Giant Chinchilla is a medium-sized rabbit, weighing 3.5–5.5kg (7lb 11oz–12lb 2oz). The powerful, broad body is somewhat elongated. The legs and neck are both fairly short and muscular.

The heads are strongly developed, especially with the bucks. The nose has a curved profile and the cheeks are substantial. The Giant Chinchilla has sturdy ears that are 150mm (6in) long on average.

Coat

The standard for the length of the Giant Chinchilla's coat varies from country to country. German Giant Chinchillas have fairly short hair but in Britain it is considerably longer. Other countries sit somewhere in between.

Colours

This breed gains its name from the colour of its coat, which is similar to the small South America rodent of that name. Giant Chinchillas have a white base coat with a clear black ticking.

Giant Chinchilla

The ticking is irregular and causes some patches or "stains" in the coat. The belly is always white. The hair nearest the belly is the undercolour.

With Giant Chinchillas, this is blue, even under the white of the belly.

Special remarks

Standardisation of Chinchillas at an international level has never really got off the ground.

For example, the Swiss do not have Chinchillas and Giant Chinchillas, they have a Chinchilla that is half way between the two.

In the USA, in addition to the "normal" Chinchilla and Giant Chinchilla, they also have the American Giant Chinchilla, which was developed by Ed Stahl of Missouri, who crossed Giant Chinchillas with Flemish Giants.

New Zealand White is an American breed.

New Zealand White/Blue/Black

Origins

The name of the New Zealand rabbit causes quite some confusion, because contrary to appearances, it is a breed that was originally bred in the United States. The New Zealand White, Blue, and Black have nothing to do with the New Zealand Red. It is thought the name stems from the use of wild New Zealand rabbits. The first New Zealand Whites were bred in 1916 by W. S. Preshaw of Rippon, California. Preshaw was not a rabbit breeder who was to be found every weekend at a show. His objective was to develop a valuable rabbit for the meat and fur trade.

This meant that in addition to being well built, the texture and quality of the pelts was important. It was intended that these new rabbits would grow rapidly so that they attained their slaughter weight quickly and the does should be fertile and produce their young without difficulty.
Which breeds of rabbits Preshaw used is unknown. Experts consider that Angoras must have played some role. Preshaw's efforts were not in vain. He had soon developed the New Zealand White that became a very successful rabbit widely used by commercial breeders.

Since there is little in common between commercial rabbit breeders and rabbit

New Zealand White

209

The New Zealand Red as bred outside the European mainland.

The New Zealand Red as bred outside the European mainland.

New Zealand Red buck that is a wonderful deep russet.

New Zealand White

fanciers who breed to show, it was quite some time before this new breed found its way to the enthusiasts.

The breed found its way to Britain after World War II, where it became popular both with breeders and laboratories. The breed was not recognised in Germany by their national rabbit breeders organisation until 1963.
At first, only the white types were recognised but the New Zealand Blacks were developed in Britain and the British were also behind the Blue types, which are virtually unheard of outside the United Kingdom.

Properties

New Zealand rabbits are generally calm, reliable rabbits, sometimes to the point of being phlegmatic.
The does are good mothers and can produce large litters. The growth of these rabbits is rapid, which is a characteristic of virtually all the breeds that were established for the table.

The New Zealand Whites are the most popular of the three types with rabbit fanciers. The New Zealand rabbit is almost unheard of as a household pet.

External characteristics

New Zealand rabbits weigh 4–5kg (8lb 13oz–11lb). They have a slightly elongated but broad and muscular body. The hindquarters are full and well rounded.

New Zealand Red buck

hairs and a dark blue undercolour. The New Zealand Blue has a blue coat and blue eyes.

Special remarks

New Zealand blues are unknown at present outside the United Kingdom and New Zealand Blacks are not yet common on the international scene.

New Zealand Red

Origins

The New Zealand Red also originates from California where it too was developed originally for the meat and fur trade for large scale production. Despite these similarities, this breed has nothing further in common with the other New Zealand rabbits. These are not the same breed.

The biggest difference between them is the weight. The New Zealand Red is clearly lighter than the New Zealand White and its derivatives. The breeds used to create this unequalled russet-coloured coat are unknown. It is surmised that the colour is derived from the Belgian Hare and that the Flemish Giant played its part too. The breed name is explained by the use of imported wild rabbits from New Zealand in the breeding.

The New Zealand Red was first seen at an American show in 1910. The first reports of exports to Europe date from 1919. In its early days, the breed was somewhat more yellow than red but the present coat has been developed by the work of various breeders to improve it.

The legs are short and powerful, as is the neck. The thick ears are about 110mm (4¼in) long on average. New Zealand Whites and Blacks have a noticeably broad head with substantial cheeks, which are particularly noticeable with the bucks.

Coat

New Zealand rabbits have normal length hair. The coat is densely haired with a considerable undercoat. The outer hairs feel somewhat coarser than other breeds.

Colours

This New Zealand breed is solely bred in white, black, and blue. The New Zealand White is internationally by far the most popular of the three and it should have a pure white coat without any markings of any kind.

The undercolour is also white, and the eyes are pink. New Zealand Blacks have dark brown eyes with lustrous dark black outer

Properties

New Zealand Reds are fairly calm rabbits but they are certainly more lively than New Zealand Whites. These rabbits have been bred commercially for their meat and pelts for some considerable time. Currently these rabbits are increasingly to be seen at shows

where they are highly regarded for both their body shape and colour.

External appearance

The New Zealand Red weighs 3–4.5kg (6lb 10oz–9lb 14oz). The bodies are somewhat elongated with muscular fore and hind-quarters.

The hindquarters are well rounded and the legs are short and sturdy. The head is quite large and with bucks in particular the head is broad with substantial cheeks. The ears are about 120mm (4¾in) on aver-age.

Coat

The hair of the New Zealand Red's coat is dense and of normal length. It feels soft to the touch and has a considerable amount of undercoat.

Colour

New Zealand Reds have a plain deep russet coat without any ticking or white hairs. The undercolour, next to the skin, should ideally be the same colour as the outer hairs but is often slightly lighter.

Special remarks

The elongated and sturdy New Zealand Red that is bred on the European mainland is significantly different to the type to be found in Britain and the United States. In these countries, the breed is more elongated, with a slightly shorter coat, and longer head.
The origins of the Belgian Hare are clearly visible in both the body and shape of the head.

Burgundy Yellow

Origins

This breed, originally from south-east France, has been pure bred for hundreds of years in France. The breed, with its characteristic yellow coat, was principally kept for the table with its pelt as an afterthought. Subsequently the fur industry started to breed selectively to improve the pelt and its colour.

By the beginning of the twentieth century, breeders in other countries started to show an interest in the Burgundy Yellow and animals were exported. It is quite astounding that an animal with a recorded pedigree that stretches back as far as that of the Burgundy Yellow is at present not officially recognised in so many countries. The standardisation of the breed between countries is also not happening.

Many countries are now deviating from the original colour, which is becoming increasingly more red as a result of cross-breeding with other breeds.

Properties

The Burgundy Yellow is rarely kept as a household pet, mainly due it its size and weight. Among those who breed to improve the line and for fun, there is a small but loyal band of enthusiasts for this breed. The breed continues to be highly regarded as a breed for the table.

The breed is virtually unheard of outside of France and the countries surrounding it.

External characteristics

The rabbits of this breed are sturdily built creatures with fully fleshed and broad

Burgundy Yellow buck

The Burgundy Yellow is a long established rabbit for the table

Burgundy Yellow buck

hindquarters. Their weight is 3.5–5kg (7lb 11oz–11lb).

The legs are fairly short, straight, and muscular. Seen from the side, the build of these rabbits appears mid-way between elongated and stocky. The Burgundy Yellow has a large head that is broad between the eyes.

The cheeks of the bucks in particular are extremely noticeable. The ears are sturdy, in any even they must not be thin, and they measure about 140mm (5½in) on average. The eyes are brown.

Coat

The coat of the Burgundy Yellow is longer than normal. Because of the substantial undercoat, it feels very soft. Judges are impressed by examples with a fine coat.

Colour

This rabbit is now being bred in other countries with a somewhat redder coat than the pedigree maintained over hundreds of years.

In any event, judges look for a uniform colour where preferably the undercoat is the same colour as the outer hair. A perfect show specimen will also have the same shade of colour throughout its coat without nuances.

Striking features are the almost white rings around the eyes and white lines under the jaw. The underside of the tail, insides of the legs, and entire belly are also white.

Argenté de Champagne/Large Silver Argenté Bleu, Brun, Crème

Origins

Large rabbits with silver ticking of their coats have existed in the Champagne district of France for centuries.
The Argenté de Champagne was bred on a large scale at the beginning of the eighteenth century. Rabbits with silver tips to their hairs were brought to France from India or Indo-China early in the seventeenth century by Portuguese seafarers.

Breeders were mainly interested in the unusual pelts of these animals. By the early nineteenth century, these rabbits found their way to other countries where considerable cross-breeding and selection took place with eventual divergence between the Argenté de Champagne and the Large Silvers, of German origin.

The Large Silvers original black with silver ticking was extended by breeding with Vienna Blues, Belgian Hares, Burgundy Yellows, and other breeds to extend the range of colours.
In Britain and America, the Argenté de Champagne is recognised, together with

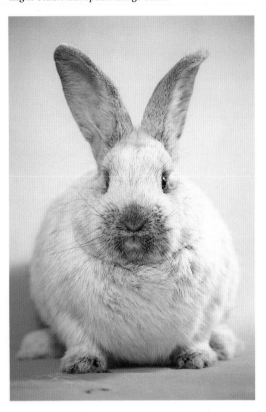

Argenté Bleu, Brun, and also somewhat rarer Crème.

Properties

These breeds are fairly calm rabbits, with a reasonably even temperament, unlike the smaller Silver. The origin of these breeds is as utility breeds, raised for the table and their pelts.
Today, the breed is largely in the hands of rabbit fanciers who spare neither time nor effort in order to breed fine rabbits with the correct colour and silver ticking. The breeds derived from the Argenté de Champagne are recognised in many countries but Large Silvers are restricted to a few.

External characteristics

These breeds weigh about 3.5kg (8lb) although the Large Silvers tend to weigh

Light black European Large Silver doe

The rabbits are born with a uniform plain colouring and start to become silver at about five to six weeks. The hairs which are shed from this age are replaced by hairs which have "silver" tips, or rather the ends of which are wholly without pigment.

The rate at which this occurs is entirely dependant on the progress of the shedding of the first hair but this is usually completed by six months.

The silver process can vary from coats that are predominantly silver to at the opposite end, coats that mainly give an impression of the colour of the base coat.

Judges prefer to see rabbits that have the same degree of silver across their entire coat, rather than those which have uneven amounts of silvering in different parts of their coats, causing dark and light patches.

Colours

The Argenté de Champagne has a black base coat with white tips and a slate blue undercoat. The Bleu is similar but has a lavender undercoat, while the Brun resembles the Bleu but is more brown and the Crème is altogether lighter.

The Large Silver is bred in rabbit grey, black, blue, brown, and yellow. They all have brown eyes except the blue types and the Large Silver browns have a rosy glow in their brown eyes.

Special remarks

The various degrees of silver are not permitted in every country. Some countries only recognise the Argenté de Champagne in light black. A rare breed derived from these rabbits is the Belgian Silver that is only bred in medium silvered black. They weigh 4–5.5kg (4lb 13oz–12lb 2oz).

somewhat heavier. The bodies are moderately elongated but certainly not long, with hindquarters that are broader than the front.

The chest is full and wide, the neck shot. The sturdy ears measure 120–140mm (4¾–5½)

Coat

The coats of these breeds are slightly longer than normal and characterised by a substantial undercoat, which makes the coat soft to the touch.

Variation of colour

Although all of these breeds have silvering in their coats, they do not necessarily appear silver. Indeed the original Argenté de Champagne are believed to have been black rabbits.

Belgian Hare

Origins

The Belgian Hare has absolutely nothing whatever to do with hares but there are still people who think that it is the result of cross-breeding with hares.

Those who understand the differences in their reproduction, development and growth will know that the differences between hares and rabbits are too great for such cross-breeding to be possible. The second part of the Belgian Hare's name does not point to its genetic origins, but to its noble appearance that resembles that of a hare.

The first part of the name is an equal misnomer since the breed was developed in Britain rather than Belgium. The progenitors of the Belgian Hare did though come from Belgium. Belgium used to regularly export

Close-up of a Large Silver

commercial rabbits for the table to Britain. These rabbits were bred for meat rather than standardised looks and it may have been that some of them had a striking reddish colour with black ticking, resembling a hare's colouring.

It is likely that enthusiasts rescued these animals from slaughter in order to breed them. There are writings that suggest that the progenitors of the Belgian Hare were Patagonians – the now extinct progenitors of the Flemish Giant.

The resulting breed is regarded by many enthusiasts as the most elegant and perfectly formed rabbit there is. The English breeder Lumb is identified as one of the first breeders of the new type of rabbit.

He showed them for the first time in 1874 and in 1887, the Belgian Hare Club was formed in Britain, indicating the appearance of the breed at shows throughout the country.

The breed was highly regarded for the table

217

Litter of black and tan Belgian Hares

Hare-coloured Belgian Hare

The Belgian Hare's popularity is not solely restricted to the keen rabbit fancier but is also kept as a household pet. Some people wrongly think the breed results from breeding with wild hares.

External appearance

Many regard the Belgian Hare as the most outstanding breed of rabbit. The body is very slim and elegant with not one coarser element to it. The average weight is 4kg (8lb 13oz) but both heavier and lighter examples exist.

The belly line of this breed is remarkable in that it runs parallel with the line of the back. The legs are long, slim, and straight with rounded feet that some liken to cats paws. Belgian Hairs have a relatively long body, with a long, thin neck and small head. The ears are also relatively long, averaging 130cm (5⅛in). They stand erect.

White Belgian Hare

but was also sought after by rabbit fanciers. Within a decade, examples had found their way to the European mainland and to the United States. The Belgian Hare became tremendously popular and fashion-able so that people were prepared to pay great sums for them.

Initially, all specimens had the "hare colouring" but European breeders crossed them with other rabbits to develop further colours.

Properties

Belgian Hares have a lively temperament and are affectionate and good-natured. The breed easily sits in the required position for the show bench assessment, although some training is required.

Belgian Hares are found in virtually every country where rabbits are bred and at shows everywhere.

Black and tan Belgian Hare

A proud Belgian Hare foster mother and her entrusted young.

Coat

The coat of the Belgian Hare is short and shiny, laying smoothly against the body with virtually no undercoat.

Colours

Litter of young Belgian Hares

HARE COLOURING

The original and still most popular colour for Belgian hares is "hare colouring". This is a reddish brown to mahogany base colour that should be as warm toned and as deep as possible, with black ticking on the ends of the hairs.

The ticking is most apparent on the back and less strong on the belly and legs. The ears have black tips and the eyes are dark brown. This colour is recognised in all countries.

The body conformation of these rabbits is identical with the normal full-coloured coat types.

Californian

Origins

BLACK AND TAN

The Belgian Hare with tan markings was first bred in Belgium and The Netherlands where Belgian Hares were crossed with rabbits bearing the genetic factor for tan markings.

These rabbits have a black back with rust brown markings. The type is mainly restricted to The Netherlands, Belgium, and a few other continental European countries.

ALBINO

Pure white Belgian Hares with pink eyes and colourless claws are almost exclusively restricted to the United Kingdom.

The Californian breed of rabbit is, not surprisingly, an American creation. Rabbit breeder George West begin the process of breeding these rabbits in southern California in 1923.

His intentions were to develop a table rabbit that also had a good quality pelt. At first, he crossed Chinchillas with Himalayans from which the unusual markings were derived. The offspring from these rabbits were in turn crossed with New Zealand Whites, which were at time the rage of the commercial rabbit breeding world.

The resulting rabbits were exhibited for the first time in 1929 but it was ten years before the American Rabbit Breeders' Association officially recognised them.

The first Californian rabbits arrived in Europe a year or so after the end of World War II and in 1958, they reached Britain. The breed was at first chiefly of interest to large scale commercial breeders of rabbits for consumption but the breed was subsequently taken up by enthusiasts.

Although the start for the Californian rabbit in the world of the rabbit fancy was a slow one, they are now to be found almost everywhere there are rabbit enthusiasts.

Californian rabbits have been described as one of the top five breeds for meat. The original colour of the nose, legs, tail, and ears of these rabbits was black but British breeders have added chocolate, blue, and lilac for the markings of the extremities, however these other colours are virtually unheard of outside the United Kingdom.

Properties

Californian rabbits are friendly and reliable animals with an outstandingly calm temperament. Because of their size, they are not much sought after as household pets. The Himalayan is a far better size for this purpose. The interest in the Californian varies from country to country but despite this, they are always to be found at international shows. The Californian is also still widely used for commercial meat production.

Californian

External characteristics

The Californian rabbit is well covered with flesh. They have broad, rounded hindquarters, a full and deep chest, and a broad

The Californian is well-known for its good nature.

Californians only have colour on their ears, nose, feet, and tail.

Three Californian rabbits

Three Californian rabbits

back. In most countries, including the USA, a slightly elongated body is required but in other countries the preference is for a shorter, sturdier build. The front is virtually as broad as the hindquarters and the head is quite sizeable with a very short neck. The average length of ear is 120mm (4¾in) and the weight varies from country to country from 3–5kg (8lb 10oz–11lb) but the average is 4–4.5kg (8lb 13 oz–9lb 15oz).

Coat

The shiny coat of the Californian is of closely packed normal length hair with a great deal of undercoat, making it dense. The hairs lays sleekly against the body.

Colour

The colour is solely born on the extremities of the body – the legs, nose, ears, and tail.

Young Vienna Blue rabbits

The rest of the body is pure white without a tinge of any other colour. Most Californians are marked with black but in Britain other colours such as blue, chocolate, and lilac have been introduced.

These are not recognised outside the United Kingdom. The eyes are without pigment, in other words pink. Californian rabbits are born entirely white but the markings develop later.

Markings

The distinctive markings of the Californian rabbit are a clear breed characteristic. The body is white but the extremities are coloured. Judges are far less strict about the shape and boundaries of the markings with this breed than they are with the Himalayan, which provided the genetic imprint for the colour markings.

Vienna Blue (and other colours)

Origins

The Vienna Blue and the other colours originate from Austria. The breed is the work of railway official Johann Constantin Schultz of Vienna-Hetzendorf who was chairman of the first Vienna rabbit breeders' association.
Johann Schultz's objective was to develop a rabbit with a fine pelt that was also first class for the meat trade. At the time blue was a highly popular colour so he set out with a blue doe of indeterminate breed, that was partially lop-eared according to accounts, plus a self yellow rabbit of rural country stock from Normandy in France, that had one ear up and one ear down, and a Flemish Giant buck, imported from Belgium. In 1895, he showed fourteen of his new breed, then known as Blue Viennese Giant, at Vienna's Prater.

The breed was officially recognised in Austria in 1897. The initial animals were somewhat heavier than today's Vienna Blues, weighing an average of 6.6kg (14lb

4oz). The colour was not then so well developed. The first Vienna Blues were exported in 1903 to Germany, The Netherlands, and Switzerland, where they quickly won the interest of many rabbit fanciers.

The breed became especially popular in Germany, so that at a large show at Hamburg in 1905, thirty Vienna Blues were entered. These rabbits weighed 6kg (13lb 4oz) on average, which was already lighter the initial stock.

A special club was founded in Germany for the breed that same year and in 1907, the breed was officially recognised in The Netherlands.
The Blue is the original and still perhaps the most loved colour but down the years, other colours have been developed by crossbreeding.

The first additional colour was white (see Vienna White), followed in by black, which was first show to the public at a large rabbit show at Leipzig in 1931.

Properties

The Vienna Blue and other coloured variants are affectionate rabbits that generally have a reasonably lively nature. The breed is one of the utility breeds that was previously bred commercially for meat and/or their pelts.

Since then, most examples of the breed are in the hands of rabbit enthusiasts. Vienna Blues are to be found in virtually every country where rabbits are bred and most of the other colours too.
The popularity for this breed is at its highest though in Austria, Belgium, Germany, The Netherlands, and Switzerland. Large classes for the breed can be found at shows in these countries.

The breed is not just popular with the rabbit enthusiasts, for in spite of their weight, the attractive appearance and excellent character has made them popular household pets.

Vienna Black

External characteristics

The coloured Vienna rabbits have a sturdy body with a broad back, and well-developed shoulders and hindquarters. The back is smoothly curved.

The strong, substantial legs are of average length. The head is also fairly substantial and the bucks in particular have very full cheeks. The ears are thick, well covered with hair, and have rounded tips. Their length should be in proportion, in practice meaning about 130mm (5⅛in) and they weigh 5kg (11lb).

Coat

The coloured Vienna rabbits coat feels soft to the touch, consists of normal length hair, and has a great deal of undercoat.

Colours

The original colour of the Vienna rabbit is plain blue with blue eyes. Subsequently, other colours have been added such as black with brown eyes, hare colour (russet base coat with black ticking and brown eyes), rabbit grey (light brown/grey base coat with black ticking and brown eyes), blue/grey (light brown base coat with blue/grey ticking and blue eyes), steel grey (steel grey base coat with black ticking and brown eyes,

and blue dun (light grey base coat, with blue ticking and blue eyes).
There were also chocolate coloured Viennas on the European mainland but these are no longer to be found.

Special remarks

The international standards for these breeds are not uniform. This means that one country breeds coloured Viennas with an elegant body and long ears, while at shows in another country entirely the opposite is preferred by judges.

The standards regarding the colours also vary widely. For instance in the United Kingdom, only the Vienna Blue is recognised and the colour range is limited in the United States.

Vienna White

Origins

Although the Vienna White bears the same name (colour aside) as the coloured Vienna rabbits, it is a quite different breed. The breed also originates from Austria, where it was developed in the early twentieth century by Wilhelm Mucke. He crossed Vienna Blues with predominantly white Dutch rabbits with blue eyes.

He exhibited the resulting rabbits for the first time at a show in Vienna in 1907. The animals were still fairly small and the influence of the Dutch stock was quite apparent. In the subsequent years it became apparent that there was interest in a white variant of the Vienna Blue from countries close to Austria, such as Germany.

The new breed was recognised in The Netherlands in 1909 and the following year the breed by Germany. For years Vienna Blues and Whites were treated as the same breed at shows but they were separated in the 1970s because it is comparatively easier to breed white rabbits true. Breeders of white rabbits do not have to contend with subtle nuances of colour that can turn up in

weighing 3–5kg (6lb 10oz–11lb). These rabbits are substantially built with broad back and hindquarters. The body should by preference slightly elongated, not be stocky.

The legs of Vienna White's are of average length and the muscularity and build should be in proportion to the rest of the body. The head is broad between the eyes and also fairly substantial, especially with bucks which have very full cheeks. The ears average about 130mm (5^1/$_8$in) long.

Coat

The Vienna White's coat is of normal length hair that feels soft to the touch, having a great deal of undercoat.

Colour

Vienna Whites have sparkling white coats and light blue eyes.

the coat of coloured rabbits, which are so important with the coloured Vienna specimens.

Properties

Because of their sparkling white coat, light blue eyes, and friendly nature, Vienna Whites are not just kept by rabbit enthusiasts but are very popular as household pets for children. In continental Europe, the Vienna White is exceptionally popular and is to be found at virtually every show.

This breed is especially highly sought after in Germany. It is so strange then that outside the continent, the rabbit is almost unknown.

External characteristics

Vienna Whites are medium-sized rabbits

Two blue Beverens from the same litter.

Beveren

Origins

The Beveren is a long standing Belgian breed, named after the town of Beveren in the Waas area. There are no precise accounts of how the breed was founded but it is felt that the Blauwe van St Niklaas and the Dutch contributed to its blood lines. It is known that before 1900 large numbers of the breed already existed around the town of Beveren.

The first show for the breed was held at Beveren-Waas in 1902. The first Beverens arrived in Britain during World War I, where certain details of the breed were further improved by enthusiastic breeders. The first specimens were also imported at about this time by Dutch and French breeders, and subsequently, the breed arrived in the United States too. British breeders though have played the main role in developing the Beveren as we know the breed today. They also developed additional coat colours. The black types were created by breeding Beverens with non pedigree black rabbits.

Beverens were at that time called Sitkas and were shown to the public in Britain for the first time in 1919. These first Beverens were soon followed by albino (white with pink eyes) and white rabbits with blue eyes.
The first chocolate Beverens were shown in the United Kingdom in 1929. Lilac Beverens, which are also a British creation, date

This Beveren shows the typical "mandolin" shaped body of the breed.

Beveren rabbits have a lively nature.

back to the 1980s. Pointed Beverens are a further type adapted from the original breed and these are shown separately from the standard Beverens although they do not differ from the main breed except in the matter of the coat colouring.

These rabbits are bred in the same colours as the other Beverens but have silver tips to their hairs. This type was recognised in Britain in 1928 but died out through lack of interest after just a few years. Several British breeders decided to develop the breed again in the late 1980s.

Properties

Beveren rabbits normally have a fairly lively nature. They were originally bred for both their meat and pelt.

Today, the breed is mainly in the hands of rabbit fanciers but it is not found in every country.

The albino type that was bred in Britain in the early part of the twentieth century no longer exists and is not recognised in any country. In addition to the blue and white Beverens, there are also lilac, brown, and black.

These latter colours are only recognised in the United Kingdom and are virtually non existent in continental Europe and the United States.

Blanc de Hotot
(Blanc d'Hotot)

Origins

External characteristics

The Blanc de Hotot (Blanc d'Hotot) is a French breed, developed by rabbit fancier Eugenie Bernard, who is one of the few

The most striking breed feature with these rabbits is the body shape that is not known with any other breed.

Netherlands Dwarf with Blanc De Hotot markings.

The hindquarters are considerably higher than the front of the rabbit, known by fanciers as a "mandolin" type or as "pear-shaped." The hindquarters are also broader and noticeably more muscular than the front. The Beveren's head is wide between the eyes and the ears are thick with rounded ends.

They measure on average 130mm (5^1/$_8$in) and these rabbits weigh about 4kg (8lb 13oz).

Coat

The black circle around the eye must be of even width.

A Beveren's coat has copious undercoat and feels soft to the touch. The hairs are 30–40mm (1^3/$_{16}$–1^9/$_{16}$in) long.

Colour

The colour of Beveren rabbits is truly striking and cannot be compared with any other colour, except perhaps that of for-get-me-nots. The eyes are blue/grey. White Beverens have a sparkling white coat and light blue eyes.

Blanc De Hotot

women to have made a mark on the development of rabbit breeding. The breed name is derived from the village where Mde. Bernard lived, Hotot en Auge.

Although Mde. Bernard insisted that she had used "ordinary" marked rabbits, experts insist that this is not possible. It is considered that rabbits with distinct markings such as Papillons and Giant Papillons formed the basis of the blood of the Blanc De Hotot. Young are occasionally born in Blanc De Hotot litters with a rudimentary dorsal stripe or with black markings on the ears and nose, which lends support to the experts' theories.

The first rabbit of this breed was shown in Switzerland in 1912 but it was quite some time before breeders in other countries took notice. The breed was introduced to Switzerland in 1927 where it became a popular breed. German breeder Friedrich Joppich imported them in 1930 but he was virtually the only German rabbit fancier to involve himself with the breed for thirty years.

The breed was officially recognised in The Netherlands in 1940 but the breed was not discovered by the British until much later; an official standard was first accepted in 1960. The breed today is not solely based on pure bred rabbits from the stock created by Eugenie Bernard.

Blanc De Hotots have also been created as it were anew by other breeders and today's specimens have Vienna White and Argenté de Champagne blood in their veins.
Although this breed is known throughout the world, most enthusiasts for it are in continental Europe. The French name has been anglicised by English speaking enthusiasts.

Properties

The Blanc De Hotot is mainly known by people who show rabbits. The breed is regarded as one of the most difficult types to breed successfully, because the required black circle around the eye is difficult to reproduce.
So far, this breed is not widely popular as a household pet; that role has been fulfilled by the Netherlands Dwarf types with similar eye circles. The rabbits of this breed are generally of a friendly nature.

External characteristics

This extremely attractive looking rabbit is characterised by its slightly elongated body with broad front and hindquarters, and broad back. Their legs are relatively short and the ears average about 130mm (5^1/$_8$in) in length. The weight is 3.5–5kg (7lb 11oz–11lb).

Coat

The coat, of normal length hair, should be densely formed, copious and springy, with a pronounced sheen and have a considerable undercoat.

Colour

Rabbits of this breed are pure white with dark brown eyes and black circles around

Blue Silver Fox

Black Silver Fox

the eyes. The pigment of the eye lids must also be dark to create the correct appearance. The claws are colourless.

Markings

The eye catching black circle markings around the eyes of the Blanc De Hotot are its major breed feature. The thickness of these should be 3–5mm ($^1/_8$–$^3/_{16}$in) and as even as possible.

Silver Fox

Origins

Silver Fox rabbits were developed more or less simultaneously in Britain, the United States, and Germany in the 1920s but Britain is considered to have played the key role in their development. It was in Britain in 1924 that the first reports were made of a "Silver Fox" rabbit with a longer than usual length coat that was marked rather like the Tans, except that what would be red was white.

The length of the coat was longer than with Tans and is more akin to the coat of Chinchilla rabbits, which is understandable, since Tans and Chinchillas were crossed with each other to produce this breed. Dominant genes suppressed what would otherwise be yellow pigmentation so that all the hairs that would have been yellow from the Tan rabbits are silvery white instead. The influence from the Chinchilla has produced the longer coat.

The breed was shown in public in Britain for the first time in 1926. Following this introduction, the Silver Fox found enthusiasts in other countries. The breed's name comes from the eponymous species of fox, which was a very popular fur animal at that time. Subsequently, breeders have also introduced Angora blood into the Silver Fox in order to improve the coat. This led to a denser coat with far more of the desirable undercoat. The introduction of Angora bloodlines has the consequence that long-haired rabbits are occasionally born to two Silver Fox rabbits. There were other colours almost from the very start of breeding these rabbits. The black was the first but this was soon followed by havana brown, blue, and lilac. The black Silver Fox is still the most popular, probably because of the greater contrast in colours.

Black Silver Fox

Properties

Silver Fox rabbits have a reasonably lively temperament. The breed is found at shows throughout the world but these rabbits are not really know by the general public. The principal attraction of the Silver Fox is its luxuriant, soft, and springy coat.

External characteristics

Silver Fox rabbits weigh 3–4kg (6lb 10oz–8lb 13oz). These rabbits are sturdy creatures with short muscular legs.

The head is fairly large, especially with bucks, with a broad skull and substantial cheeks. The ears are well covered with hair and stand upright, measuring about 120mm (4¾in) on average.

Coat

The shiny coat of the Silver Fox rabbit is about 30-40mm (1³/₁₆–1⁹/₁₆in) long, which is longer than most breeds. The coat is dense with a considerable amount of undercoat.

Colour marking

The belly, toes, a ring around the neck, lips, and nose of these rabbits have silver/white markings. There is also a silver/white ring around the eyes, and the insides of the ears, plus the underside of the tail which are similarly silver/white. Finally there is a triangle of silver/white on the neck behind the ears.
There are also hairs that are noticeably longer, which have silver/white tips on the chest, shoulders, sides, legs, and upper side of the tail.

Colours

Silver Fox rabbits are bred in a number of colours. The most widespread ones are the blacks, which like the brown types always have brown eyes. The blue rabbits have blue eyes as do the blue sables, while the brown sables have brown ones. The brown Silver Fox rabbits and both of the sable varieties have eyes that glow red when seen from certain angles.

Harlequin (Japanese)

Origins

This breed started out life as a Japanese, yet it has nothing whatever to do with the Orient and is actually a French breed.
The original rabbits that founded the breed were found by accident in a park by a rabbit fancier where semi wild rabbits and tricolour Dutch rabbits were probably kept. The colouring was so unusual that the enthusiast decided to establish a new breed. The first rabbits from this new breed were shown in 1887 and also entered in the World

Harlequin

Harlequin

Typical zebra stripes of a Harlequin

Most Harlequins are not marked as preferred for the show ring.

rabbits with wild rabbit colouring. Initially rabbits with this "wild colour factor" and those without it were acceptable but subsequently only specimens without this factor were accepted.

The breed is in the hands of rabbit fanciers from its outset and the breed was never popular as a meat or pelt provider. Although the origins of the present Harlequins virtually all stem from the original French rabbits, other countries had also produced a harlequin-type rabbit but these were never elevated to the status of a breed.

Properties

Harlequins have a fairly normally lively nature and are generally friendly. The breed is far more popular in some countries than others. The breed is not that popular as a household pet, due largely to lack of awareness of the breed but also to its size and weight.

Exhibition of 1889 in Paris as Lapin Japonais (Japanese rabbit). Breeders from other countries became interested in the quite unusual rabbit and offspring were busily exported. The change of name in Britain and the United States is related to World War II. Many of the early offspring clearly had the genetic factor for the wild rabbit colouring, with a lightly coloured belly, light circles around the eyes, and other hallmarks of

The great difficulty in breeding ideally marked specimens makes the Harlequin an enormous challenge. Breeding two finely marked rabbits does not guarantee equally good offspring and those that have sub-standard markings can easily produce first-class young. This characteristic of the breed has meant that not every enthusiast remains with the breed because of the potential for disappointments.

External characteristics

Harlequin rabbits weigh 2.5–4kg (5lb 8oz–8lb 13oz), making them medium-sized specimens. Harlequins are sturdy rabbits with a slightly elongated body, and legs of average length, which are muscular. The head is large, particularly with bucks and the ears, which are rounded at their tips are about 120mm (4¾in) long.

Coat

Harlequin rabbits have a short, shiny coat with hair that lays sleekly with an average extent of undercoat.

Colour

Harlequins are two coloured rabbits that occur in four different colour combinations. One of those colours is always yellow to orange while the contrasting colour can be black, havana brown, blue, or lilac.

Blue and lilac Harlequins have blue/grey eyes and both brown and lilac examples have eyes that glow fiery red when seen at a certain angle. The black types are still the most usual, not least because the other colours are not recognised in most countries.

The belly, underside of the tail, and insides of the legs are usually lighter in tone. The eyes (with the exceptions given above) are dark brown. Harlequins sometimes have white hairs on their belly, beneath the tail, and the insides of their legs.
This is not treated as a major fault by show judges.

Markings

The breed standard for Harlequins requires sharply defined boundaries between the colours, without patches of the opposite colour with the markings.
The perfect example has a two-colour head with the line between the colours exactly in the centre. Beneath the black portion of the head, the chest is yellow to orange and the leg is black; beneath the yellow to orange half of the head, the chest is black and the leg yellow to orange. It is acceptable for the chest and leg to be the same colour, pro-vided this contrasts with the colour on that side of the head.

Additionally, the ear on the yellow/orange part of the head should be black and vice versa. The body should have patches of the two colours resembling zebra stripes. The number is not important, but they should be clearly defined and if there is a change of colour between one side of the rabbit and the other, it should be in the centre of the back. This standard of perfection is almost impossible to achieve and show judges mark accordingly.

Magpie

Origins

Magpie rabbits are really Harlequins with which the yellow/orange patches are white. The removal of the colour in a rabbit can be achieved by cross-breeding with rabbits with the "Chinchilla factor", such as Silver Fox or Chinchilla.

The dominant gene of these breeds impedes the formation of the yellow pigmentation. This means that everything that would be yellow/orange with the factor is instead silver white.

Magpie rabbits are a recent arrival on the scene and originate from the United King-dom.
The name is derived from the similarity of colour combination with a magpie and the original black and white of the original Magpie rabbits.

Magpie rabbit with typical opposed markings

Magpie rabbit

Young Magpie rabbit

Properties

Magpie rabbits are fairly calm and even tempered. Because there can be so many disappointments in breeding Magpie rabbits, few breeders gamble on it, since breeding two perfectly marked rabbits with each other does not guarantee the offspring will be similarly marked.

Luck plays a major role in the breeding of Magpie rabbits. The breed is not very popular as a household pet, mainly due to the lack of awareness but also because of these rabbits size.

External characteristics

Magpie rabbits weigh 2.5–4kg (4lb 13oz–8lb 13oz). Their bodies are slightly elongated but the neck is very short. The bodies are also well rounded and broad with short muscular legs.

The heads are quite sizeable, especially with the bucks, with substantial cheeks. The ears have well rounded tips and are about 120mm (4¾in) long.

Coat

The hair in a Magpie rabbit's coat is short, dense, and shiny. These rabbits have a normal amount of undercoat.

Colour

Magpie rabbits are two coloured animals that like the Harlequin are bred in four colour combinations. One colour is always silver-white: the contrasting colours can be black, havana brown, blue, or lilac. The black and brown rabbits have brown eyes; the blue ones have blue/grey eyes. The eyes of the brown and lilac rabbits can glow fiery red when seen from a certain angle.

The black and white Magpie rabbits are the most popular so far, perhaps because the contrast between black and white is the greatest, which is what makes these rabbits so unusual. The fact that the other colours are not recognised in most countries has a bearing too, of course, but the interest is not very great in those countries where they are recognised.

Markings

The same standards apply to the Magpie rabbit as with Harlequins. With Magpie rabbits it is also almost impossible to achieve the perfect markings according to the standard.

In reality, these rabbits often have a substantial amount of white hairs in the black parts of the body and vice versa, and the colour segments are not always very straight or sharply defined.

Special remarks

Germany has its own Magpie rabbit, known as the German Rhône. Unlike Magpie rabbits, there are no precise specifications for the markings.
The standard merely requires the coat pattern to resemble the trunk of a birch. German Rhônes are also slightly smaller than Magpie rabbits, weighing 2–3kg (4lb 6oz–6lb 10oz).

Rhinelander

Origins

The Rhinelander is a German breed that like the Papillons has striking markings on a predominantly white body. The markings are on the head and sides and they also have a coloured dorsal stripe.
The breed was developed at the start of the twentieth century by postman and rabbit fancier, J. Heintz of Grevenbroich, in North Rhine-Westphalia, which explains the choice of name. The German postman used non pedigree rabbits with rabbit grey coats with patches and Harlequins.
The first offspring that he produced that resemble today's Rhinelanders had wild

Blue and black Rhinelander does

Blue Rhinelander buck

Two young Rhinelanders

Young Rhinelanders

rabbit coloured coats with yellow patches. Subsequently, the stock was crossed with Papillons to improve the markings of the Rhinelander and to achieve uniform black patches for greater contrast. It was this contrast which attracted many enthusiasts to the breed. The breed was officially recognised in Germany in 1905 and then found its way to other countries. At first these rabbits were imported by enthusiasts from surrounding countries The breed was not discovered by breeders in other countries until much later and did not reach Britain and the United States until the 1950s.

It has never achieved the same level of popularity in these latter countries that it enjoys in continental Europe. The white Rhinelander with black and yellow patches was for a long time the only variety bred and shown. The blue variety is a more recent addition, which was only recognised in the 1980s.

Properties

Rhinelanders are generally friendly rabbits with a normally lively temperament. The breed is not solely kept by rabbit fanciers, with many being kept as household pets for both children and adults.

External characteristics

Rhinelanders weigh 3–4kg (6lb 10oz–8lb 13oz), placing them in the medium-sized category of rabbits. These rabbits are fairly thickset and rounded, with a quite short neck.

The length of the legs should be in proportion with the body. These animals – particularly the bucks – have strong, round, and wide heads. The ears stand upright and are about 120mm (4¾in) long.

Coat

Rhinelanders have normal (short-haired) coats that lay sleekly smooth against their bodies with a marvellous shine. The coat feels soft to the touch.

Colours

There are two varieties of Rhinelander: white with black and red–yellow markings and dark brown eyes, and white with blue and yellow markings and blue eyes.

Markings

The striking characteristic of this breed is the unusual colour markings on an otherwise sparkling white coat. A differentiation is made between the head and body markings. The markings on the head include the "butterfly", "thorn", circles around the eyes, cheek markings, and coloured ears. The "butterfly" is a dark patch on the nose from one corner of the mouth to the other that is more or less the shape of a butterfly. This should be sharply defined. The "thorn" is a mark precisely in the centre of the nose. The circles around the eyes are also required markings that should be of even thickness around each eye. The cheek markings and fully coloured ears are also laid down in the breed standard. There is a dorsal strip along the back, running from behind the ears to the tail. Finally, these rabbits also have various markings on each side which should ideally be round and not run into each other. The perfectly marked specimen has two of these markings.

Special remarks

There are a number of special points about the breeding of Rhinelanders. Virtually every litter has specimens that look like badly marked Harlequin rabbits, and almost every litter has rabbits that are entirely white.
The offspring often have a dark "moustache" of black or blue which some say is where the name "Charlie" (after Charlie Chaplin) came for, although the term is used in English to indicate any incorrectly marked or coloured specimen.

Thüringer

Origins

It is not surprising in view of the name to learn that the Thüringer is a German breed of rabbit. The breed originates from the village of Waltershausen in the Thüringer Wald in eastern Germany next to the border with the Czech Republic.

The breed was developed there late in the nineteenth century by a local schoolmaster named David Gärtner, who also bred some

Thüringer

This Thüringer buck has a distinctly well-developed head.

Thüringer colouring is not developed when young.

of the smaller breed of rabbit, such as Silver, and Himalayan.

By cross-breeding Giant Papillons that had no markings with his Himalayans and Silvers, he sought to achieve larger Himalayans and Silvers but there was little interest in the specimens that resulted and they were not recognised.

Gärtner did not let this deter him and he continued with breeding using the amber coloured rabbits that had resulted from this breeding and managed to establish this unusual colour. He called the new rabbits "Chamois" and these rabbits were far more successful.

The breed was recognised in Germany in 1907. The original Chamois rabbits were not precisely like today's Thüringers.
These rabbits weighed about 2.5kg (5lb 8oz). By cross-breeding with Flemish Giants and careful selection, the Thüringer rabbit as we know it today was eventually created.

Properties

Thüringer rabbits are friendly and have a normal lively nature. The breed is popular with rabbit fanciers in continental Europe where it can be seen at virtually every show. The breed is known in Britain and the United States but has a very small following.

External characteristics

Thüringer rabbits should be somewhat stockily and fully built with fairly straight, sturdy legs. The head should be substantial and broad. The ears with this breed are about 120mm (4¾in) long and Thüringers weigh 3–4kg (6lb 10oz–8lb 13oz) or more.

Coat

The hair in the coat should be of fairly fine texture and densely formed, laying sleekly flat to create a marvellous sheen.

Colours and markings

Thüringers are only bred in the one colour. Judges pay close attention to the correct colouring and markings. These rabbits are amber with a darker tip to each hair that gives a subtle tint to the coat, which must not become too dark. The colour is darker on the ears, chest, nose, hindquarters, legs, belly, underside of the shoulders, and flanks than the rest of the body and should be much darker on the nose, ears, and belly. By

ciers, the breed has not found much interest elsewhere. The first specimens were imported into the United Kingdom in the 1990s and the breed was recognised by the British Rabbit Council in 1994.

Properties

Sallanders are normally active rabbits that have not yet caught on as household pets. In The Netherlands, they have a sizeable group of interested fanciers.

External characteristics

Sallanders are somewhat stockily but well built with a short neck. The legs are of normal length but must be solid and sturdy. The head of this breed is substantial and broad with sturdy ears that are about 120mm (4¾in) that are carried erect. These rabbits weigh 3–4kg (6lb 10oz–8lb 13oz).

Coat

The coat is of normal length hair that is dense with a substantial undercoat. The hairs should feel soft and lay sleekly against the body.

Colours

The colour of the Sallander is not found with any other rabbit. It is an attractive and

blowing on the hair it is possible to see that the colour of the hairs near the roots is much lighter. Thüringer rabbits have dark brown eyes.

Sallander

Origins

The Sallander or Salland rabbit as it is also known, has similar colouring and markings as a Thüringer but much paler. The breed is of Dutch origin and is named after the Salland area of The Netherlands where the breeder lived. The first examples were bred by a well-known Dutch rabbit judge D. J. Kuiper from Olst in the province of Overijssel who crossed Thüringers with Chinchillas. The resulting new breed was recognised in The Netherlands in 1975 where the breed is well-known from its appearances at shows. With the exception of British fan-

Small breeds

English/Papillon/English Butterfly/English Spot

Origins

The English, Papillon, English Butterfly, or English Spot is one of the most eye-catching and popular breeds of rabbit and in spite of the French alternative name, the breed originated in Britain, as did so many other breeds.

The breed was developed in the middle of the nineteenth century by selecting spotted rabbits of non pedigree stock that bore a resemblance to the Great Lorrainese (subsequently known in Britain as Giant Papillon). Right from the outset, the breeders objective was to create an attractive rabbit with an unusual coat, which was rare for those times, when the main objective was usually to produce a good commercial rabbit

Young black Papillon

Rabbit grey English buck

bleached out version of the Thüringers coat. In common with Thüringers, judges examine the colour and markings carefully at shows.

The off-white hairs all have brown/black tips which casts a tint across the body. The ears, chest, nose, hindquarters, legs, and belly are darker and the undersides of the shoulders are also darker than the rest of the body.

The darker colour is most clearly to be seen on the nose, ears, and belly. The eyes are dark brown.

with excellent meat and or pelt. The new breed was very popular in Britain between 1850 and 1860 but then became neglected. to reappear twenty years later, this time to stay.

The first examples were almost solely with black markings. Later, tri-colour, rabbit grey, blue, and chocolate were also bred. The first pedigree Papillons arrived on the European continent in 1889, where they first met with enthusiasm among German fan-

A British English

Blue Papillon

Papillon with dark amber markings

ciers. However, at this time only commercial fur or meat breeds were judged at shows. The Papillon was never intended for this purpose and consequently breeders had great difficulty to maintain the new breed. Fortunately these rules did not apply to shows in the surrounding countries where Papillons slowly began to appear with greater regularity.

Many new colours were developed, so that Papillons are recognised in far more colours in continental Europe than the English in Britain or English Butterflies in the United States, where the colours are restricted to the original black, rabbit grey, tri-colour, blue, and chocolate.

International standards for this breed have never been successfully established for this breed. This manifests itself in different physical shape, colours, and markings, which can vary from country to country.

The name Papillon, meaning butterfly, is derived from the butterfly-shaped marking on the nose, that is also found with other breeds (including the Giant Papillon). In the United States, this breed is usually known as English Butterfly. The English has been an important in the development of other breeds of rabbit.

English/Papillon markings show early: this rabbit is a few days old.

Properties

The English and related breeds is popular with breeders but also much sought after as a household pet for children because of their compact size and fascinating markings. These rabbits are best described as friendly, attentive, and lively.

Breeding English/Papillons is not an easy matter because mating two of them together produces a litter with half the offspring being butterfly marked, a quarter self coloured rabbits and a further quarter of plain white rabbits. (See also special remarks).

External characteristics

These are slender rabbits with a long body that shows off the required markings to best effect.

The neck is normally elongated and clearly visible. The sturdy legs should be in proportion with the body. The ears are about 110mm (4¼in) long and these rabbits weigh 2.5–3.5kg (5lb 8oz–7lb 11oz).

Coat

The English/Papillon has a short, shiny coat that is dense and lays sleekly against the body and feels soft to the touch.

Colours

These rabbits are bred in a wide variety of colours but not every colour is recognised in each country. The most common colours are black and blue but there are also rabbit grey (light grey/brown base coat with black ticking), steel grey (light grey base coat with black ticking), blue/grey) light grey/brown base coat with blue/grey ticking), havana brown, amber, and cream dun. In addition, there are also examples that are tri-coloured, having a white base coat with markings in two colours (like the Rhinelander). Some countries treat these tri-colour specimens as a separate breed at shows.

The colours of the eyes are derived from the colour of the markings, so that black, rabbit grey, steel grey, and amber rabbits have brown eyes, and the blue, blue/grey, and cream dun ones have blue or blue/grey eyes.

Markings

The markings on the body and head are an important feature with these rabbits. Opinions on what are the correct markings vary from country to country but in virtually every country, good examples have an unbroken dorsal stripe that is straight and sharply defined. The standards for markings on the sides of these rabbits vary very wide-

Blue/grey Papillon buck

ly: in the Netherlands for example they should be the size of fresh peas but in other countries they may be much larger. Although the opinions on body markings do vary widely, there is general agreement internationally about the markings of the head. These include the butterfly, a "thorn" markings, cheek flashes, rings around the eye, and coloured ears.

The butterfly is a butterfly-shaped marking on the nose that should be clearly defined and run from one corner of the mouth to the other. The "thorn" is a marking that should be precisely in the centre of the top of the nose.
The rings around the eyes are mandatory markings that should be of even thickness. The standards also lay down markings on both cheeks and fully coloured ears.

Special remarks

Breeding of these rabbits has its ups and downs. The required markings are difficult to breed with any certainty which means that every litter not only includes correctly marked specimens but also some that are self-coloured (one plain colour), and others that are all white.
Sometimes the white rabbits have a small dark "moustache" on their nose which some believe is the origin of the term "Charlies" used to describe them but the term is more widely used for any incorrectly colour or marking on a rabbit. In German speaking countries, these rabbits are called *Weiszlinge*.

Havana

Origins

Dark brown rabbits with white patches were born from two non pedigree rabbits belonging to a Dutch rabbit fancier called Honders at the end of the nineteenth century.

This colour was at that time extremely unusual. Thanks to the efforts and inspiration

Young Havana

a Mrs Illingworth imported the first Havana rabbits into Britain. Finally, the rabbits made their entrance into the United States in 1916.

From the very start, the exceptionally colour of these rabbits was equally sought after by the fur trade and rabbit en-thusiasts. Subsequently, this exceptional rabbit has become common place in virtually every country where rabbits are bred.

Many new breeds have been created with blood lines from the Havana, such as Perlfee, Alaska, Marburger, Thrianta, and Lilac.

Properties

Havana rabbits are chiefly kept by rabbit breeders who wish to keep the breed established and to improve it. Few Havanas are kept by people as household pets, though they are entirely suitable for this purpose.

These animals have a lively and attentive nature without being nervous.

External characteristics

Havana rabbits are stockily built, with a short body, which is broad and well rounded at the front as well as at the hindquarters. Little neck is visible with these rabbits, the head is short and very substantial and broad – especially with the bucks.

The ears are rounded at their ends and measure about 110mm (4¼in). The legs are of average length. Havanas can weigh 2.5–4kg (5lb 8oz–8lb 13oz).

Coat

The coat of an Havana has an average extent of undercoat and it is dense, and sleek. The texture of the hairs is fine.

Colours

The colour is the major breed feature of Havana rabbits. This should be a plain dark

of the breeders van der Horst, Muysert, and Jacobs, this colour was not lost for all time. By crossing with Himalayans and then crossing back with the original brown buck, they attempted to establish the colour, and thankfully they succeeded. The resulting rabbits were shown in 1902 at a major international exhibition in Paris. Because of the fiery glow in the eyes of these rabbits, their initial Dutch name referred to their "fiery eyes" and then later they were given another Dutch name but eventually they became known as Havana.

Those Dutch breeders were not the only ones busy creating chocolate brown rabbits. Similar rabbits were also developed in France by crossing non pedigree and Himalayan rabbits. Many of these early Havana rabbits unfortunately had white patches in their coats – an inheritance from their original parents – but through carefully selection, this now only occurs infrequently.

Once the rabbits had been shown at further exhibitions, breeders in other countries became interested in this exceptional colour. It is known that the first Havana rabbits reached Switzerland in 1905, and were introduced into Germany in 1907. In 1908,

Havana

Two-week old Alaska rabbits

chocolate brown, without nuances. The preference is for specimens where the colour runs as far back towards the tail as possible. The under colour is blue, preferably as dark as possible.

Judges prefer rabbits at shows that have no white hairs at all and where the chocolate colour tends towards red or grey it is regard-

ed as a fault. The same is true of rabbits that are so dark that they appear at first to be black.

A typical characteristic of this breed is the fiery red glow in the eyes when viewed from a certain angle.

Alaska buck

Alaska

Origins

This breed originates from Germany, where it was developed by rabbit judge Max Fischer from Gotha and breeder Schmidt from Langensalza.

The name Alaska seems a strange one for a black rabbit because it conjures up snow scenes and white rather than black rabbits. The idea behind the name was the original intention to create a breed of black rabbit with white hair tips, rather like the Alaskan fox.

The breeders crossed Silvers, Dutch rabbits, Havanas, and Himalayans resulting in two types of offspring: plain black rabbits and others with longer coats with white points. There was interest in both type from other breeders but in spite of their efforts, the

Head of an Alaska rabbit

Alaska rabbit

breeders were unable to create a rabbit with a coat resembling that of an Alaskan fox. They decided to settle for a pitch black coat and subsequently they bred their stock with plain black Papillons.

The breed was shown for the first time in 1907 and several years later, stock was exported to other countries. The Alaska rabbit eventually came to be present almost throughout the world.

Properties

Alaska rabbits are lively creatures that are predominantly kept by those who breed for a challenge. These are the serious fanciers who strive to maintain and improve the breed. Few of these rabbits are kept as household pets.

External characteristics

The body type of the Alaska is similar to the Havana. The Alaska rabbit is also stocky, without any discernible neck and with a well rounded body.

The head is well developed, broad, but not long. The ears are rounded at the ends and measure about 110mm (4¼in). The breed is of medium size, weighing 2.5–4kg (5lb 8oz–8lb 13oz).

Coat

The coat of the Alaska consists of fine hairs that feel soft to the touch, with a normal amount of undercoat.
The hair lays sleekly.

Colours

Alaska rabbits are pitch black over their entire body. The belly and chest can be slightly less darkly coloured. These rabbits have dark brown eyes.

Marburger

Origins

This German breed with is unusual blue coat was developed by one of the few women breeders in the history of development of rabbit breeds, Frau Sandemann who started her quest in 1916 by crossing Vienna Blues with Havanas.

The offspring from these unions she in turn crossed with light coloured black Silvers. The resulting breed was recognised in Germany in 1920 and is name after Marburg, where Frau Sandemann lived and bred her rabbits. Marburgers subsequently played an important role in the development of Lux rabbits.

The colour is genetically the same as the English Lilac or Dutch Gouwenaar but

This fine Marburger is shown to perfection.

because Marburger breeders have constantly selected breeding stock from darker specimens, the colour of the German breed is now different from the coats of these other breeds.

Properties

Marburger rabbits have a lively and friendly nature. The breed has a set group of enthusiasts in mainland Europe but is virtually unheard of elsewhere.

External characteristics

The physical shape of Marburger rabbits is very similar to Havanas.
Marburgers are short, rather squarely built rabbits with broad chests and gently rounded hindquarters with a short muscular neck and broad but short head. The legs are in proportion to the body in terms of both

their length and thickness. The ears are sturdy with rounded tips, measuring about 110mm (4¼in). These rabbits weigh 2–3.5kg (4lb 6oz–7lb 11oz).

Coat

The soft-feeling coat consists of short, springy hairs that have a sheen. The hair lays sleekly smooth.

Colours

Marburger rabbits are blue with a subtle brown haze. Their eyes are blue/grey but appear fiery red when seen from a certain angle.

Special remarks

The rabbits of this breed do not hide their origins: the influence of Silvers makes itself apparent from time to time when offspring are born that develop silver tips on their hair after a time.

Perlfee

Origins

This breed with its exceptional colour originates from Germany. It was developed early in the twentieth century by several breeders, including K. Hoffmans of Dusseldorf who was also responsible for developing the Lux breed. His original objective was to breed a rabbit that resembled the coat of a Siberian squirrel because these were much in demand by furriers.

He named his creation Düsseldorfer Perlfee. At that same time, another breeder from Augsburg named Deiniger crossed Havanas with agouti rabbits to achieve a very similar result.
His breed he named Augsburger Perlfee. These two breeds eventually came together

in the one breed known as Perlfee. The breed is mainly restricted to Germany and bordering countries but a small number of fanciers in Britain breed and show this rabbit.

Properties

There is healthy interest in the Perlfee in continental Europe. The breed has a friendly, good-natured, and lively temperament.

External characteristics

The build of the Perlfee is similar to the Havana. These rabbits are short, somewhat squarely built, with broad chest, shoulders, and back, with gently rounded hindquarters.

The neck is very short and muscular and the legs are in proportion to the body. The head is substantial, broad, but not long with ears that have rounded tips, that measure 110mm (4¼in). These rabbits weigh 2–3.5kg (4lb 6oz–7lb 11oz).

Coat

The lustrous coat feels soft and springy to the touch. The short hairs are densely formed and lay sleekly smooth.

Young Perlfee rabbits

Colour

The Perlfee's colour is unique. This is a blue grey rabbit with changeable colours on the tips of the hairs. This pearl effect is unique and is only found on the tips of the hairs.

The German name is derived from this pearl effect. The belly, parts of the legs, and the underneath of the tail are white, together with rings around the eyes, the nose, a line around the neck, and the insides of the ears. The eyes are blue/grey.

Lux

Origins

The Lux was developed by the same Dusseldorf rabbit breeder who was responsible for the Perlfee. Mr Hoffmans used Perlfees, Marburgers, Tans, and Sables for his creation.

The Lux was first shown in Germany in 1919 and recognised there in 1922 and given its name. The breed was exported to various countries in Europe and is now represented in virtually al the continental European countries but is quite unknown in Britain and the United States.

Properties

The Lux is a rabbit that is wholly bred for showing and almost exclusively kept by

Coat

The coat feels soft and springy and is short-haired, shiny, and sleek.

Colours

The colour of the Lux rabbit is quite unusual. The undercolour is pure white, the base coat is warm orange, and the tips of the hairs are a delicate silver/blue.

The belly, parts of the legs, and the underside of the tail are white, together with circles around the eyes, the nose, a line around the neck, and the insides of the ears. The undercolour on the white parts of the body is blue. The eyes are blue/grey but glow fiery red when viewed from certain angles as a result of red light being reflected by the ball of the eye.

The Lux has a soft pastel-like coloured coat

rabbit enthusiasts who strive to improve the breed and to assess their efforts in front of a judge on the show bench. The breed would be perfectly suitable as a household pet for both adults and children.

The fact that few do is mainly due to lack of awareness of this breed among the general public.

External characteristics

The build of this rabbit is similar to that of the Havana. These rabbits are squarely built, short, with broad chest, shoulders, and back, and with gently rounded hindquarters.

The neck is very short and muscular and the legs are in proportion with the body. The head is substantial and broad – especially with the bucks – and must not be too long. The ears are sturdy with rounded ends and they measure about 110mm (4¼in).

These rabbits weigh 2–3.5kg (4lb 6oz–7lb 11oz).

Special remarks

An American breed that strongly resembles the Lux is the Palomino. This United States breed has a different background to the Lux. The Palomino was bred by Mark Youngs of Washington and was recognised in USA in 1952 in two different varieties: the Lynx and Golden Palomino.

The Lynx is the variety that most resembles the Lux. These American rabbits weigh 4–5kg (8lb 13oz–11lb). These American rabbits are virtually unknown outside the USA.

Lilac/Gouwenaar

Origins

There are three breeds of rabbit that closely resemble each other and that share the same genetic background: the English Lilac, the Dutch Gouwenaar, and the German Marburger. Although the German breed is now much darker than these other two breeds, it originally shared the same soft pastel blue of the Lilac and Gouwenaar. The first breeder to achieve this remarkable colour was the British breeder H. Onslow of Cambridge who exhibited his Lilac rabbits, as he called them, at a major show in London in 1913.

The British woman breeder, Mrs Illingworth, who brought the first Havanas to Britain in 1908, also bred this colour of rabbit, which she called Essex Lavender. Somewhat later in the early 1920s, the geneticist Prof. R. Punnet of Cambridge bred a similar colour from Beverens and Havanas, which he called Cambridge Blue.
These breeds were subsequently merged in Britain into one breed, now known as Lilac.

Independently of these English breeders, the Marburger was bred in Germany with a very similar colour.

This breed, which has sub-sequently been bred to be much darker, was recognised in Germany in 1920.

Gouwenaar

Gouwenaar

249

The first Dutch lilac coloured rabbits were bred by a pigeon breeder called Spruyt who also bred Havanas as a hobby. He made no serious effort to breed lilac rabbits; lilac sports appeared spontaneously from his Havanas.

The new colour was named after the famous Dutch cheese town of Gouda and the name Gouwenaar is widely used on the continent for the offspring from Mr Spruyt's original lilac rabbits and as a colour name for other European lilac rabbits.

Gouwenaars were recognised as a new breed in The Netherlands in 1927.

Properties

Both the English lilacs and Dutch Gouwenaars are calm, affectionate creatures with a gentle nature. The breeds are not enormously popular but are bred by a select band of rabbit enthusiasts. The breeds would make first class household pets but few of them are kept as such. This is probably due to the lack of knowledge about them on the part of the general public.

External characteristics

The build of both Lilac and Gouwenaar rabbits is similar to Havanas, so that these are squarely built, short rabbits with broad chests, back, and shoulders and gently rounded hindquarters. The short, broad head is carried on a very short, muscular neck. The legs are in proportion with the body. The ears are sturdy, with rounded ends, and measure about 110mm (4¼in). Rabbits of both breeds weigh in the range 2.5–3.5kg (5lb 8oz–7lb 11oz).

Coat

The hair is fine and soft, short, and lays sleekly against the body.

Colours

Both the Lilac and Gouwenaar have pastel-like lilac coats. The eyes are blue/grey but when viewed from a certain angle, the eyes glow fiery red.

Beige

Origins

Rabbits with this unusual colour were bred in both Britain and The Netherlands and breeders in the two countries have bred these rabbits under the same name without each being aware of the other.

The Beige arose in Britain towards the end of the 1920s but lost favour after a few years to the many different coat colours that are bred in Britain. The Dutch Beige was developed in Rotterdam in the 1930s by a breeder, Mr G. Brinks, who also named the breed.

The breed was recognised in The Netherlands on May 1 1940, a few days before

Beige

Beige

World War II came to The Netherlands. About forty years later, British rabbit fanciers discovered this breed in The Netherlands and introduced it to their country, where it became known as Isabella.

Properties

Beige rabbits are virtually unknown outside of Britain and The Netherlands, where they are almost exclusively in the hands of rabbit fanciers.

These rabbits show themselves to be generally pliable and reliable rabbits with a gentle nature.

External characteristics

The build of these rabbits is similar to that of the Havana. They are squarely built, short, with broad chest, shoulders, and back, and with gently rounded hindquarters.

The neck is very short and muscular and the legs are in proportion with the body. The head is substantial, broad and short. The ears are sturdy with rounded ends and they measure about 110mm (4¼in).

These rabbits weigh 2–3.5kg (4lb 6oz–7lb 11oz).

Coat

The hair is fine and soft, short, and lays sleekly against the body.

Colours

These rabbits have a beige coloured coat but part of the coat consists of light pastel blue tips to the hair. The eyes are blue/grey but appear fiery red when viewed from certain angles.

Steenkonijn/ "Stone rabbit"

Origins

The Steenkonijn is a Belgian breed of long standing. The Flemish name "Steenkonijn" literally meaning stone rabbit is derived from a former Belgian weight unit of "a stone" which was about 3.5kg (7lb 11oz) or roughly equivalent to one British "stone" in weight.

This was the slaughter weight of these rabbits that were shipped in huge quantities to Britain for the table but once the British started to import rabbit meat from Australia, the interest in the breed dwindled in Belgium. By the end of the nineteenth and start of the twentieth century, the breed had all but died out. A breeder named Delounois thought this was a great shame and set to, with the help of other breeds, to build the breed up again.

He found a buck at a show that looked much like the near extinct Steenkonijn and bred it with a doe that had some resemblance to the

original standard for the breed. By careful selection, he managed to recreate the breed.

He showed his first examples of the breed in 1932 and he won official recognition for the breed on June 12, 1934. Mr Delounois was certainly the most important breeder of "Stone rabbits," which he bred for more than forty years.

In spite of his efforts, the breed never became widely popular, despite rising interest in Belgium. There is virtually no breeding of these rabbits outside Belgium.

Properties

Most examples of this breed are lively, inquisitive, and good natured rabbits.

These rabbits are mainly in the hands of Belgian rabbit fanciers but their convenient size and character would make them ideal pets for children or adults.

The Belgian Steenkonijn is lively and good natured.

External characteristics

These lively and friendly rabbits weigh 2–3kg (4lb 6oz–6lb 10oz) with an ideal weight of 2.5kg (5lb 8oz). Their bodies are stocky with short necks and a substantial broad head.
The legs are short and muscular. The wide ears stand upright on the head and are on average 100mm (4in) long.

Coat

The hairs of the coat are fairly short, dense, and lay sleekly smooth against the body.

Colours

The Belgian Steenkonijn ("Stone rabbit") is bred in three wild agouti colours: rabbit

Young "Stone rabbits"

grey (light grey/brown base coat with black ticking and brown eyes), hare coloured (russet base coat with black ticking and brown eyes), and steel grey (grey base coat with black ticking and brown eyes).

This latter colour is recognised but is fairly rare.

Deilenaar

Origins

The Deilenaar is a creation of Dutch breeder G. W. A. Ridderhof of Deil in the fruit-growing Betuwe area of The Netherlands. Which rabbits he used to create this breed is not known but experts guess that Flemish Giant, Chinchilla, New Zealand Red, and Tan blood flows through their veins. There is even a suggestion that the Belgian Hare may have been involved. The breed was recognised immediately before the outbreak of war in Holland on May 1 1940. It is perhaps not surprising therefore that its spread to other countries was somewhat delayed.

The breed was only recognised in the United Kingdom at the end of the 1980s. The Deilenaar is bred in a limited way in a number of European countries but is barely found elsewhere.

Properties

The Deilenaar is a strong, robust breed of rabbit that is normally lively and friendly. The breed in principally in the hands of breeders who breed for pleasure and a challenge and has not yet become widely kept as a pet although they are certainly suitable to be household pets.

External characteristics

The Deilenaar is a rugged, sturdily built animal that is strong and muscular but with a short body that is broad and a short, thick neck and imposing head.

The legs are in proportion with the body and the ears are about 110mm (4¼in).
These rabbits weigh 2–3.5kg (4lb 6oz–7lb 11oz).

Young "Stone rabbits"

Deilenaar

Head of a Deilenaar buck

Young Deilenaars

Coat

The spotted ticking of the coat with its red tint is the most eye catching characteristic of the Deilenaar rabbit.

The coat is longer than many other rabbits, has an average amount of undercoat, and a marvellous sheen. The hair is shorter on the head, ears, and legs than the rest of the body.

Colours

Deilenaars have a warm red/brown base coat with an irregular and fairly heavy black ticking that is spread over the entire body. Agouti rabbits normally have lighter areas in the insides of their ears, rings around the eyes, lips, nose, and also the underside of the tail, belly, and insides of the legs are lighter coloured and free from ticking.

These two Deilenaars are good examples of the patchy dark ticking.

Siamese Sable

Special remarks

The French have their Brun Marron de Lorraine which is virtually identical in colour to the Deilenaar but the build of the two rabbits is quite different. The Brun de Marron de Lorraine is a slender rabbit with an angular head and a short-haired coat that weighs 1.5–2.5kg (3lb 5oz–5lb 8oz).

Sable/Marten Sable

Origins

Sepia/brown Sable

In common with other breeds, the Sable was developed almost simultaneously in several countries.

The British were probably first, with the results of cross-breeding by breeder D. W. Irving of Himalayans and Chinchillas, at the beginning of the twentieth century. These rabbits had darker extremities such as the nose, ears, tail, and legs, with a lighter body colour but with less contrast than with Himalayans.

The resulting rabbits were shown in 1920 but were not recognised. Subsequently this type of Sable, known in the United Kingdom as Siamese Sable and elsewhere as a Marten, has become a well-established sight at shows.

Another breeder who developed similar rabbits was Emil Thomsen from Stellingen in Germany who used a number of breeds, including Chinchillas, blue Havanas,

Viennas, Belgian Hares, and Thüringers. He named his creation "Stellinger Kaninchen" or Stellingen rabbits. A French breeder named M. Fraineau from the Cognac region also developed similar rabbits using Chinchillas and albino rabbits which he showed at a major international exhibition in Paris in 1925. Finally there are also reports of American breeders developing similar rabbits at this same time. The first Sable rabbits were various shades of sepia/brown but blue was subsequently developed in Germany and yellow or Siamese is of much more recent origin. This latter colour was developed during the 1940s by a well-known German rabbit judge called Friedrich Joppich. At about the same time a variety was developed in Britain and the United States which is virtually unknown in continental Europe.

This was a rabbit with similar white markings as a Silver Fox. This type of rabbit was initially known in Britain as a Maraaka but this was changed to Marten Sable and in the USA these rabbits are termed Silver Marten or Silver Marten Sable.

The name Marten refers to small mammal related to the sable family, which have similar colourings as these rabbits.

Seen from a certain angle, Sable rabbits have a fiery red glow in their eyes.

Medium blue Sable

Properties

Sable rabbits are lively creatures that generally have a friendly nature. The breed was originally principally bred for the fur trade but now enjoys interest from those who breed as a hobby.

Because of the interesting colourings, convenient size, and their friendly nature, these rabbits are also widely kept as pets for both children and adults.

External characteristics

Sable rabbits weigh in the region of 2.5–3.5kg (5lb 8oz–7lb 11oz). These rabbits have a slightly elongated body that is well fleshed out.

The powerful legs are of normal length. The head of a buck should be substantially built and in common with other breeds, the doe's head is smaller. The ears are about 110mm (4¼in) on average.

257

Coat

Sables/Marten Sables are bred in a variety of colours, including blue, sepia/brown, and Siamese (or yellow). Colours other than sepia/brown are virtually unheard of outside Europe.

The colours can be in three different shades: dark sable, medium sable, and light sable. The sepia/brown and Siamese (yellow) rabbits have brown eyes, and the blue types have blue/grey eyes. All colours have a fiery glow in their eyes when viewed from a certain angle.

Colour markings

There are two types of Sable rabbit. The rabbits illustrated are known as Siamese Sables in Britain and USA, and as Martens or Marters in continental Europe.

These are rabbits with darker colouring of the extremities than the rest of their body. The other type is the Marten Sable or Silver Marten which have the same colourings as the other type but with the addition of white colour markings similar to the Silver Fox.

Special remarks

Breeding of Sables/Martens has its idiosyncrasies. Mating two medium toned Sables will almost always produce a litter with medium-toned offspring but also some darker specimens, together with white with colour markings, and albino rabbits. Mating an albino with a darker Sable will produce all medium coloured offspring. Darker coloured Sables paired with each other always produce similar young.

Silver

Origins

The Silver is a British breed and quite possibly one of the oldest breeds of rabbit in the world.

There are written sources which contend that rabbits with silver tips to the hairs of their coat were brought back to England from Asia during the heydays for merchant adventurers. These rabbits were said to have been kept and bred by Buddhist monks long

Light black Silver gaining its silver coat

Head of a yellow Silver buck

Light-blue Silver during silvering

before the birth of Christ. Other sources disagree with this Asian origin for these rabbits and suggest that the silver-tipped coat arose spontaneously as a sport or mutation among rabbits kept in man-made warrens in Lincolnshire in the middle of the nineteenth century. These rabbits became known as Lincolnshire Sprigs, Lincolnshire Silver Greys, or Millers.

This story overlooks the fact that rabbits with silver tips to their hair were recorded in England in the seventeenth century by the writings of Gervase Markham in 1631. A remarkable point is that this writer talks of Riche rabbits, since the old French name for silvered rabbits was Lapin Riche. What became of these rabbits is not entirely clear. What is known is that attempts were being made in the middle of the nineteenth century to develop Silvers as a pure breed. The breed was first shown in 1860 and the breed had found its way to most countries in Europe by the middle of the nineteenth century although it would be some years before it was recognised.

The first English breed standard for Silver rabbits was set up in 1880. The breed was above all important for the fur trade and good breeding stock was exported for large sums of money.

The breed arrive in the United States before the end of the nineteenth century. At that time, the only colour was black, which were called silver/greys but these were soon followed by yellow. It has always been considered that this colour originated from France where Burgundy Yellow rabbits were crossed with Argenté de Champagne rabbits, to produce the cream-coloured type also known as Argenté crème. These Argenté crème rabbits were probably imported by British breeders to develop the yellow type of Silver. There are sources though that dispute this theory and suggest that English fawn-coloured rabbits were used to produce the yellow type of Silver. A further colour is the rabbit grey Silver.

These were bred by an Englishman named G. Johnson of Kettering. According to

Medium black Silver

Medium brown Silver

The nose becomes silver first

reports, he used Belgian Hares with Silvers and called his creations "Brown/Silver." The other colours of blue and brown were probably created by crossing with Vienna Blues and Havanas.

The introduction of a new colour in an existing breed can happen in all manner of ways and it is not impossible to imagine that

in the long family tree of the Silvers – much of which is unknown – many other breeds and non pedigree rabbits may have been present.

Blue Silvers and brown Silvers were developed on the continent, rather than in Britain and blue was only recognised in the United Kingdom in 1980 but remains rare in both Britain and the USA.

The brown variety of Silver is virtually unknown in Britain and the United States, while it has quite reasonable popularity in continental Europe.

Properties

This breed of small rabbit with silver tips to the hairs of its coat has a lively and prickly temperament. This temperament makes these rabbits less suitable for keeping as children's pets. Silvers are represented in virtually every land where pedigree rabbits are kept and bred.

Medium yellow Silver

Medium black Silver

These rabbits are popular show animals even though it is quite difficult to breed them according to the breed standard for the extent and uniformity of the silvering of their coat.

External characteristics

Silvers are small rabbits that weigh in the range 1.75–3.25kg (3lb 14oz–7lb 3oz). These rabbits have a stocky build with a short body and very short neck. The chest is broad and the hindquarters are well rounded. The broad head with full cheeks is quite striking and this is seen at its best with the bucks. The legs, which are straight and muscular, are in proportion to the body. The ears are wide with rounded ends; their length varies from 80–120mm (3⅛–4¾in).

Coat

Silvers have short, densely formed hair in their coat, which lays sleekly smooth and

has a high sheen. This lustre is most pronounced with the darker and medium-toned colours.

Varieties

The Silver gets its name from the silvering in its coat which appears gradually. These rabbits are born a plain self colour but their coats begin to get silver hair tips around five to six weeks.

All the hairs that are then shed are replaced with ones that have silver tips. In reality, the tips of the hair are wholly without pigment. The speed with which this happens is entirely dependant on the rate at which a rabbit sheds its hair but it usually takes until the rabbit is about six months old.
Silvering occurs in three different levels of intensity. Light is when most of the body is silvered so that little of the base coat colour beneath remains visible, dark occurs when

Light rabbit grey Silver doe with young

only part of the coat is silvered so that the appearance is of a coloured rabbit, and medium is between these two extremes. Judges at shows prefer rabbits that have uniform silvering, without any darker and lighter patches.

These categories of colour shade are not universal for all countries.

Colours

Independent of whether the silvering is light, dark, or medium, the colour of the coat can be rabbit grey, black, blue, brown, and yellow.

The colour of the eyes is determined by the colour of the coat, so that black, yellow, and rabbit grey specimens have brown eyes, while blue types have blue eyes but brown rabbits have brown eyes that can glow fiery red.

Halle Pearl-Grey

Origins

The precise origins of this breed – known in Belgium as Parelgrijze van Halle – are unknown but it is generally considered that it comes from the Belgian rabbit breeder Vervoort of Halle who discovered rabbits with this unusual colouring in a litter of

The Halle Pearl-Grey is a Belgian breed.

Halle Pearl-Grey

Havanas. A point of conflict still unresolved is the colour of the eyes of this breed. The first examples of this breed had brown eyes according to reports and articles and the breed standard still stipulates brown eyes but this is genetically impossible.

All the rabbits of this breed have blue/grey eyes and it is not possible to breed this colour coat with any other eye colour. The breed has a group of enthusiasts in Belgium but is virtually unknown elsewhere.

Properties

The Halle Pearl-Grey is a compact breed of rabbit with a fairly docile and friendly temperament.

The breed is principally kept by serious enthusiasts but considering its small size, winning ways, and unusual colouring, it would make a first class household pet for children.

External characteristics

The Halle Pearl-Grey has a stocky, short body that is well fleshed out. The neck is very short and the head is broad, rounded, and short with cheeks that are particularly pronounced with the bucks. The ears are about 90mm (3½in) long and they stand upright. These rabbits weigh 2–2.5kg (4lb 6oz–5lb 8oz).

Coat

The coat is of fine, normal-length hair with an average amount of undercoat.

Colour

The name of this rabbit is based on its unusual colour which is a very soft blue/grey with a light sheen; the belly is somewhat more matt. These rabbits have blue/grey eyes regardless of the standard requiring brown ones.

Chinchilla

Origins

Chinchilla rabbits originated in France where they were very popular long before the Giant Chinchilla (or Chinchilla Giganta) was bred. The Chinchilla is in reality an agouti rabbit from which all the yellow or red pigment has disappeared. The factor that is responsible for this change is genetically dominant.

The name of the breed is derived from the South American chinchilla that was very popular with the fur trade for many years. In common with that South American rodent, the Chinchilla rabbit was originally bred for the fur trade. This was not just because of the similarity with the pelt of a chinchilla, the Chinchilla rabbit's coat matured so quickly that it could be skinned at five months. This is much later with other rabbits, so the Chinchilla was an attractive commercial proposition for breeders. The breed was originally

Two young Chinchilla rabbits

Chinchilla rabbits are inquisitive and friendly creatures.

because of lack of knowledge. The breed is very well known among those who show and breed rabbits throughout the world.

External characteristics

Chinchilla rabbits weigh 2–3kg (4lb 6oz–6lb 10oz). Their build is short, stocky and broad with strong, short legs and neck. The head is also short and broad and quite substantial with bucks.

The ears are sturdy and about 100mm (4in) long, with rounded tips.

Coat

The coat of a Chinchilla rabbit is longer than most other rabbits, although not as long as that of Angora and Silver Fox. The hairs are about 30–40mm ($1^3/_{16}$–$1^9/_{16}$in) long and are

Chinchilla

bred by an engineer, M. J. Dybowski. He utilised Himalayans, self-coloured blues, and agouti rabbits. The first Chinchilla rabbits were exhibited at Saint-Maur in France in 1913 and after they were shown at a major international exhibition in Paris in 1914, breeders from other countries became interested in the breed.

The first Chinchilla rabbits reached Britain the following year and Germany several years later. The breed was discovered by the Americans in 1919 and quickly became extremely popular.
The Chinchilla contrib-uted to the development of other breeds, such as the Silver Fox, Sallander, and Sable/Marten.

Properties

Chinchilla rabbits are friendly creatures with a normally lively nature. This small sized breed is not widely kept as a household pet

Thriantas are lively and inquisitive.

Properties

Thrianta rabbits have a small but enthusiastic group of breeders that support the breed. Given the small size and friendly nature, these rabbits make ideal pets, although few of them are kept as pets because the breed is not widely known.

External appearance

Thriantas are small and stockily built rabbits that weigh 2–3kg (4lb 6oz–6lb 10oz). Their backs are quite short and they are broad chested with sturdy, straight legs of an average length. The hindquarters should have a fine rounded line. Thriantas have a well-devel-oped broad head that is rounded but short. This head is more pronounced with bucks. Their ears are rounded at the tips and are about 80–100mm ($3^1/_8$–4in) long.

Coat

The thick undercoat makes the coat soft to the touch. The hair is of normal length but densely formed.

Colour

The colour of Thriantas is a reddish orange which should be as deep a tone as possible. In common with most breeds, the hair on the belly and beneath the tail is lighter.

Common faults with this breed are a pale colour, white belly, and dark rims to the ears. Ticking in the coat, however slight, is undesirable. The eyes are dark brown.

Sachsengold (Saxon Gold)

Origins

Sachsengold rabbits are very similar to Thriantas, although their colouring is slightly less warm toned. The breed was reared by German breeder Nennack from Röhrsdorf. Tan, Havana, Chinchilla, Silver, Harlequin, New Zealand Red rabbits, plus perhaps non pedigree rabbits with yellow colouring, were used in the development of this breed with its evenly coloured plain yellow coat.

Sachsengolds were first shown in Leipzig in the former East Germany in 1953. The breed is mainly bred in Germany and neighbouring countries, although not in The Netherlands where breeders prefer to raise Thriantas.

Properties

Many Sachsengold rabbits are also kept at children's pets, as well as being bred and shown by enthusiasts. These rabbits have a lively and friendly nature.

External characteristics

Sachsengold rabbits have a slightly stocky build with well rounded hindquarters. Their legs are straight and strong, the necks are short, and the head is broad and short. These rabbits have relatively short ears that are well covered with hair. They weigh 2–3.5kg (4lb 6oz–7lb 11oz).

Coat

The coat consists of normal length hair that has an ample undercoat, which makes it feel soft to the touch.

Colour

A yellow to orange colouring with a lighter coloured undercoat is preferred. The eyes are dark brown.

Hulstlander

Origins

The Hulstlander is a Dutch breed of recent origins which gets its name from the district in the Province of Overijssel where the breed was developed.

Breeder J. de Graaf used Vienna Whites and blue-eyed Polish rabbits. The breed was developed in the early 1980s and was accepted in The Netherlands in 1984. At present the breed appears to be entirely restricted to the country of its birth.

Properties

This small rabbit with its sparkling white coat and pale blue eyes is kept and bred on a limited scale by a small group of en-

Hulstlanders are active and temperamental

thusiasts. The breed is too temperamental to be kept as a pet for children.

External characteristics

The Hulstlander is a small rabbit with a squat build. The body is short and plump with well rounded shape. The short legs are sturdy and straight.

The neck is very short with a broad, short head and substantial cheeks. Seen from profile, the head is very rounded. The ears general are about 90mm (3½in) long. These are sturdy and are carried in a "Vee" shape. Hulstlanders weigh about 2.5kg (5lb 8oz).

Coat

The short-haired coat of a Hulstlander is shiny, lays sleekly smooth, and has an adequate undercoat.

Colours

Hulstlanders are only bred white. The coat is sparkling white, without any markings or yellow tinge, and the eyes are pale blue. Some people call rabbits without any pigmentation in their coat and pale blue eyes "luminous."
The Hulstlander is similar genetically in terms of colour with the Vienna White, blue-eyed Polish, and Angora rabbit.

Tan

Origins

This popular breed was developed in the second half of the nineteenth century in England.
The first Tan rabbit was a sport that was discovered by a Mr Cox of Brailsford who kept a mixture of rabbits, including Dutch and Silvers but also cross-bred and non-pedigree rabbits running freely. The Tan, with its black guard hairs and pale yellow belly, attracted attention from a number of breeders who decided to establish the sport as a breed.

They called the rabbits "Black and Tan." The original Tans had yellow belly with black on

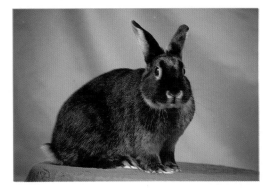

original Black and Tans. This resulted in two types existing: the Cox bred "Brailsford" type that was fairly small, stocky, and temperamental, and the "Cheltenham" type developed by enthusiasts that was larger and with a quieter nature.

Two separate breed societies were formed: the National Black and Tan Club, founded in 1890, which promoted the "Brailsford" type and the British Black and Tan Rabbit Club which brought together enthusiasts for the type of rabbit bred by Purnell in Cheltenham. The National Black and Tan Club was eventually disbanded and the Black and Tans were united in one breed. Today's Tan is a combination of the two types: it has the colour of the Cheltenham type and the build of the Brailsford type.

Interest was shown in the breed by enthusiasts in other countries around the turn of the nineteenth century into the twentieth and a number of rabbits were exported. Originally, in addition to the black type of Tan rabbit, there were also specimens with blue in their coats that were developed by

their backs and had stocky bodies. The British breeder Purnell of Cheltenham crossed Black and Tans with Belgian Hares in order to redden the yellow belly colour to a deeper tan.

The influence of the Belgian Hare blood provided both the desired darker colour on the belly and a slimmer, longer body and the new cross-breeds were also larger than the

Tan rabbit

Brown Tan buck

English fancier Atkinson. Within a few years, the brown type of Tan also came into existence. Later again in 1927, a lilac coated Tan was developed by cross-breeding Tans with Marburgers. Tan rabbits have in turn had enormous influence in the development of breeds such as Thrianta and Silver Fox.

Properties

The Tan has its enthusiasts in every country where pure-bred rabbits are kept. These rabbits have a friendly nature and are widely kept as pets, in part because of their convenient size.

External characteristics

Tan rabbits have a stocky, solid build with well rounded form. Their legs are straight and sturdy but in proportion to the body.

Head of a Tan buck

Two young Tan rabbits

The red triangle behind the ears can be seen.

The neck is very short with a large head that is very short and has substantial cheeks. Does have a somewhat smaller head. The ears are sturdy with nicely rounded tips and are about 90mm (3½in) long. Tan rabbits generally weigh 2–3kg (4lb 6oz–6lb 10oz) or more, making them small rabbits.

Coat

The coat is of average length with fine hairs that lay sleekly smooth and have a deep sheen.

Colours

The most usual and original colour is black (with tan), followed by blue, and brown. The most recent addition is the lilac Tan. Black and brown Tans have brown eyes, blue and lilac Tans have blue eyes.

Both the brown and lilac varieties have a fiery glow in the eye which can be seen from certain angles. This occurs in quite a number of breeds. The tan markings should be a warm reddish brown.

Tans often have a few white hairs in their coat but too many are considered a fault for a show specimen.

Colour markings

The colour markings ideally should be clearly defined with sharp edges to form a contrast between the black, blue, brown, or lilac of the main colour.

The tan colouring appears around the eyes, on the front and insides of the ears, the nostrils, beneath the jaw, chest and belly, and the insides and back of the hind legs. There are also small patches of tan on the feet. A stripe of tan colouring runs around

Black Klein Lotharinger

A lilac Tan

the neck, and along the edge of the jaw, ending in a triangle of tan behind the ears. Finally, there are also points of tan from the belly to about half way along the flanks.

Klein Lotharinger (Dutch Papillon)

Origins

The Klein Lotharinger is a smaller Dutch version of the Giant Papillon or Great Lorrainese (Lotharinger is German and Dutch for Lorrainese). The breed was first recognised in The Netherlands in 1975 but became fairly popular within a short space of time.
The breed was developed by a female rabbit breeder and judge called J. M. K. Berman van Schelven of Emmeloord.

She wanted to develop a smaller version of the Giant Papillon, which she achieved by cross-breeding Netherlands Dwarfs with Giant Papillons. In a later stage, Papillons (English/American Checkered) rabbits with coarse markings were also used to improve the markings.

Properties

The Klein Lotharinger is fairly lively but good natured. The breed is mainly kept by breeders and enthusiasts who show them but they certainly make good household pets. The breed has become extremely popular in The Netherlands but similar types of rabbit are also popular elsewhere with local versions. The Germans have their Kleinschecken and the Czechs also have an own variant. Elsewhere this type of rabbit is represented by the English or Papillon or American Checkered.

External appearance

The Klein Lotharinger has a slightly elongated body that is generally to be found with all marked breeds, developed because the longer body displays the markings to better effect.

The legs are straight and sturdy, the head is broad, not very long, but well developed. The bucks in particular have substantial cheeks. The ears are about 100mm (4in) and these rabbits weigh about 3kg (6lb 10oz).

Coat

These rabbits have normal length hair in their coat which feels soft to the touch. The coat is densely formed and lays sleekly smooth with a strong sheen.

Markings

Markings against a sparkling white coat are characteristic of this breed and the other related breeds. The body and head markings

Blue Klein Lotharinger

Klein Lotharingers are active and friendly.

Colours

Klein Lotharingers have colour markings in virtually all the known rabbit colours. The most usual colour is black, because it provides the best contrast but other colours such as blue and havana (brown) are fairly common.

There are also examples with rabbit grey (light grey/brown base coat with black ticking), steel grey (light grey base coat with black ticking), blue/grey (light grey/brown base coat with blue/grey ticking), blue dun (light grey base coat with blue ticking), amber, and greyish/yellow.

Tri-coloured Lotharingers are also recognised. The colour of the eyes depends on the coat colour: black, brown, rabbit grey, steel grey, and amber coloured examples have brown eyes, and blue, blue/grey, blue dun, and beige rabbits have blue eyes.

Greyish/yellow Klein Lotharinger

are treated separately in the breed standard. The head markings include the "butterfly", circles around the eyes, "thorn," cheek slashes, and coloured ears. The "butterfly" is a dark patch shaped liked a butterfly on the rabbit's nose.

This should be clearly defined and stretch from one corner of the mouth to the other. The eye rings are also mandatory markings, which should be of equal thickness. The "thorn" is a marking above the centre of the nose.

The cheek markings and fully coloured ears are also required in the standard. Along the back of these rabbits there is a dorsal stripe from being the ears to the tail. This should ideally be of even thickness and clearly defined.

Finally, these rabbits have markings on their sides which should ideally be round and clearly separated.

White hairs in the coloured markings is regarded as a fault in a show specimen.

Brown Klein Lotharinger

known markings were known hundreds of years ago and paintings from the fifteenth century show rabbits with Dutch markings, though not as finely marked as today's Dutch rabbits.

Surprisingly, in view of the name of the breed, this is not a Dutch breed but an English one although the ancestors of the breeding stock came from The Netherlands and northern Belgium where large numbers of grey rabbits with white chests and throats were bred under the name of Brabander in the early nineteenth century, for sale to the meat trade.

These rabbits are said to have weighed 2.5–3kg (5lb 8oz–6lb 10oz). Many of these rabbits were exported to Britain where they were known as Dutch rabbits.

Rabbit fanciers bred some of these rabbits to develop the Dutch rabbit as we know it today, with a white blaze, white front, and white hind legs.

Black Dutch rabbit

Special remarks

It is quite difficult to breed Klein Lotharingers true in view of the fact that two perfect specimens cannot be guaranteed to produce equally perfect offspring. Almost every litter includes plain coloured rabbits and some that are wholly white as well as others that are marked.

A broken "butterfly" marking resembles a moustache and this may be the origin of the nickname "Charlies" given to wrongly marked or deviating colour specimens. Some suggest it is because these "moustaches" look like that of Charlie Chaplin.

Dutch

Origins

The Dutch is one of the oldest breeds there is, alongside the Silver and Tan. The well-

Young black Dutch rabbits

The breed was already being shown in 1870 and nine years later a club was formed for the breed.

The first of these rabbits to be exported were sent to Germany in 1882. At that time, the majority of the rabbits had black markings that had been created by selection from grey and white rabbits. Blue and brown were soon added, together with virtually every agouti colour such as rabbit grey and steel grey.

Little is recorded about how the colours were arrived at although it is known that these colours arose spontaneously among pedigree and non pedigree rabbits in various countries at about this time. It is unknown which country or breeder was responsible for which colours. The origins of one of the latest colours is certainly known.

The first tri-colour Dutch were bred in The Netherlands by a breeder known as Vijlbrief, who first showed them to the public in 1922. Subsequently breeders Versteeg and Wassink also bred three-coloured Dutch rabbits with unusual markings that were achieved by crossing with Harlequins (then known as

Three young blue Dutch rabbits

Japanese). These rabbits were exhibited at a major show in Rotterdam in 1925.

Mr Wassink wrote a standard work in Dutch about the breed. The Tri-colour Dutch rabbits have found their way to other countries.

The colour was imported into Britain in the 1960s and has become accepted in other countries too. In both Britain and the United States, Dutch Tri-Colour are treated as a separate breed.

Dutch rabbits have played an important role in the development of other breeds, such as the Vienna White, Harlequin, Alaska, Polish, and in Germany the blue-eyed Hussemer.

Properties

Although the Dutch rabbit was derived from rabbits bred commercially for the table and for their fur, the breed itself was never intended as other than a rabbit for the fancier and they have also become widely known as pets. This is not surprising in view of their unique markings, wide choice of colours, and calm, easy-going nature.

External appearance

The Dutch is a stocky, fairly substantial looking rabbit with quite short, straight and

Brown Dutch buck

Amber coloured Dutch

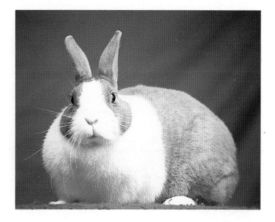

Blue/grey Dutch buck

sturdy legs. The short rounded head is particularly broad between the eyes and has very full cheeks. The ears of these rabbits are quite large with rounded tips that are about 90mm (3½in). No standards have been successfully achieved regarding the weight of Dutch rabbits. The lightest ones are in Britain where the maximum weight is 2.25kg (4lb 15oz). The upper limit is 3.25kg (7lb 3oz) in Germany and Switzerland and 2.75kg (6lb) in The Netherlands. Most Dutch rabbits weigh between 1.5–3kg (3lb 5oz–6lb 10oz).

Coat

The hairs of the Dutch rabbit's coat are short and densely formed. They also lay sleekly smooth with a strong sheen to the entire coat.

The standards for these rabbits require two coloured areas on the head that incorporate the eyes, cheeks, and ears but do not extend to the nose.

When correctly marked, the Dutch has a clear inverted V-formed blaze of white on its face. The rabbit should be coloured from about the middle of its back to the tail with a clearly defined and straight dividing line.

The feet and hind legs should be white. Judges pay special attention at shows to the clear definition of the markings. Although this breed is bred throughout the world, the standards for markings are not interna-tionally accepted and vary from country to country.

The question of whether the colour should run behind the ears or not is an area where there is not agreement.

Dutch rabbits are principally bred in black and this is also the colour most frequently seen at shows.

Indeed, this colour is so popular that many people do not even realise that Dutch rabbits are also bred in other colours. In addition to black, blue and brown Dutch rabbits are relatively common but there are also rabbit grey (light grey/brown base coat with black ticking), hare colouring (russet base coat with black ticking), steel grey (light grey base coat with black ticking), blue/grey (light grey/brown base coat with blue/grey ticking), amber, beige, chinchilla (silver/white base coat with black ticking), and orange.

In addition to this there are tri-colour Dutch rabbits. Not all colours are recognised everywhere. The colour of the eyes depends on the colour of the coat: brown eyes with black, brown, rabbit grey, russet, steel grey, chinchilla, and amber examples and blue eyes with blue, blue/grey, and beige ones.

Exceptionally fine Dutch buck in fairly uncommon rabbit grey colour.

Tri-colour Dutch were created by crossing with Harlequins

Himalayan

The tri-colour has similar colour markings to a Harlequin rabbit but combined with white Dutch rabbit markings. If a tri-colour is black or brown with yellow, then it has brown eyes; if blue with yellow, the eyes are blue.

Special remarks

It can be very difficult to breed Dutch rabbits with good markings with very big variations in every litter. From time to time Dutch rabbits are born with light blue eyes, such as those of the Vienna and Hulstlander but this colour is considered a fault. Lighter coloured patches can also occur in the eyes.

Himalayan (Russian)

Origins

The name of this breed rather suggests eastern origins, but the breed as we know it today came from Britain. Britain is not only the birthplace of so many different breeds of rabbit, but also of many rodents, particularly those with unusual coats.

Himalayan rabbits, or rather rabbits with the striking colour markings of this breed, have a long history. Rabbits of a similar type existed hundreds of years ago in China, where they were regarded as symbols of

fertility and as offerings to the gods at the start of each new year. There are written accounts of virtually white rabbits being seen in the wild in Russia that had dark coloured noses, ears, and legs.

It is not entirely clear how such rabbits found their way to Britain. Various old sources suggest that English merchant adven-turers brought them back from the

Himalayans are predominantly calm with a friendly nature.

279

Himalayas. The same story is given for the origins of Siamese cats, which have similar types of markings, although different eye colourings from Himalayan rabbits.

These differing stories have led the breed to have had a variety of names. These rabbits have been known as Polish, Chinese, Himalayan, Russian, Egyptian Smut, and even African rabbit. Various local names referred to the marking on the nose.

Whatever the early history of this breed, it was certainly bred in Britain in the middle of the nineteenth century under the name Himalayan, and that is still its name in English speaking countries today.

At about this same time, French breeders also bred a similar rabbit that according to scientists has no genetic links with the Himalayan and must therefore have had different origins.

The first Himalayans had black markings but later other colours were also produced, although black was and remains the most popular colour because of the excellent contrast it achieves.

Properties

Himalayans are predominantly calm with a friendly nature. At first the breed was mainly sought after for its pelt but these days they are mainly bred for showing.
Himalayans are to be found in virtually every country that has rabbit fanciers and at most shows. The popularity of this breed of small rabbit is not restricted to the serious enthusiast, because with their good temperament, compact size, and unusual colour markings, they make first class pets.

External characteristics

Himalayans have a fairly elongated, slender build with hindquarters that are well rounded. Their legs are fairly long and slender too. The head, which is fairly long and slim, is in keeping with the body. The ears are of average length – about 90mm (3½in).

Young Himalayan, not yet fully marked.

There is no international standard for the weight of Himalayans. In most countries this is set at about a maximum of 2.5kg (5lb 8oz) but there are countries where these rabbits weight 3kg (6lb 10oz).

Coat

Himalayan rabbits have a short-haired smooth coat that is densely formed and soft to the touch.

Colours

The colouring of Himalayan rabbits is the result of a mutation. The actual colouring of the rabbits only shows itself on the body extremities such as the ears, nose, legs, and tail which are colder through poorer circulation. The colour is imprinted through a recessive gene.

The main colour that is shown at present is black but examples with blue and brown markings can be seen. There have also been Himalayans with silvering on the marked parts of their bodies, although this strain has almost died out. Examples with brown and lilac markings are virtually restricted to Britain and the USA.

It is theoretically possible to introduce almost any colour marking by cross-breeding with other rabbits but the colour of the eyes is always pink.

Colour markings

The dark markings on Himalayan rabbit's coats only appear on the body's extremities: the ears, nose, legs, and tail. A fine show specimen has an oval marking on the nose that runs beneath the jaw. If there are any white hairs in the colour "mask" or if the markings are irregular shaped, points will be deducted by judges.

The darker colour on the ears should commence where the ears are joined to the head and like the nose markings, these should be sharply defined. Himalayan rabbits are born white or mouse grey and the dark markings only start to appear after a few weeks. It is not possible to judge the quality of the markings until the rabbit is at least six months old.

The colour is quite susceptible to change and tends to become darker when the weather is colder and lighter in the summer. If these rabbits are injured, the hair that re-grows over the scar will be dark, not white. The same thing happens if the hairs are clipped short, but this must never be done for show rabbits.

It is normal for the older rabbits to acquire dark rings around their eyes, but although this a perfectly normal occurrence that is characteristic of the breed, it is still regarded as a fault by judges.

Special remarks

There is also a larger version of the Himalayan that is bred in a few countries in Europe as a Large Russian. These rabbits look like the Californian (American variant of the Himalayan), although less sturdy and more elegant though they are still described as slightly stocky and square shaped. The Large Russian is regarded as a table rabbit and weighs 3.5–5kg (7lb 11oz–11lb).

Dwarf breeds

Polish

Origins

Polish rabbits, despite their name, are an English breed that first were bred in the middle of the nineteenth century. The rabbits probably first came from Belgium where little rabbits were bred with Dutch, Silvers, and other small non pedigree rabbits.

There are sources which suggest that a breed

Superb pink-eyed Polish white

A sable marten Polish and British Giant

Rabbit grey Polish

known as Lapin de Nicard existed in France in the eighteenth century that weighed 1.5kg (3lb 5oz); this rabbit is regarded by some as the forerunner of all dwarf breeds.

Although considered as a Fancy breed today, the Polish rabbit was considered a delicacy so that they were bred for meat in spite of their diminutive size. At that time, Polish rabbits weighed 1.5–2kg (3lb 5oz–4lb 6oz).

The British perfected these Belgian table rabbits and the breed was recognised in 1884. The foremost proponent of the breed at that time was a judge and breeder named John Meynell from Darlington England. By the end of the nineteenth century Polish rabbits had found their way to other countries.

The breed was solely reared in the albino (white with pink eyes), and white with blue eye varieties. Coloured variants were regularly born to Polish parents and during the 1950s, the British Rabbit Council

eventually decided to permit coloured examples to be shown.

Properties

Polish rabbits are lively creatures that because of their fine looks and small size are extremely popular as household pets as well as with serious rabbit fanciers.

The dwarf breed is tremendously popular in Great Britain and the United States. The breed is known as Hermelin in continental Europe but is only bred and shown on a limited scale.

External characteristics

Polish rabbits are small, compactly built, but elegant animals with slender, straight legs. The head is also slim and somewhat wedge shaped when viewed from the side, with a fairly pointed nose.
The ears are fairly small and stand upright, close together to give these rabbits an attentive appearance. Their ears are about 60mm (2⅜in) long and these rabbits weigh 700g–1.5kg (1lb 8oz–3lb 5oz).

Coat

The short-hairs of the coat are fine, smooth, and densely formed.

Colour

Polish rabbits are now bred in almost all the known rabbit colours.

Special remarks

The Polish breed is known as Britannia Petite in the United States where they also have developed the American Polish rabbit which has a different, much rounder build, which is stockier and more fleshed out than the Polish, placing it mid-way between the English Polish and Netherlands Dwarf. The

American Polish also has a rounder head with a short nose. There are separate classes for the two breeds at US shows.

Netherlands Dwarf (White)/Dwarf Polish

Origins

Netherlands Dwarf rabbits are the continental European version of the English Polish rabbit where they are known as Dwarf Polish.

The two breeds have the same origins but have grown apart through selective breeding so that they have different body conformation. The first albino Polish specimens to reach the continent arrived in Germany in the late nineteenth century and then found their way to The Netherlands. The breed became extremely popular in both countries and was recognised in The Netherlands in 1907.

In the middle of the twentieth century, both German and Dutch breeders started to develop an even smaller Polish rabbit, with dwarf body, head, and ears. There was no minimum weight stipulated in the breed standard at that time and the de facto norm was "the smaller the better," leading to judges preferring the smallest and lightest examples.

Because breeders consequently only bred their smallest rabbits, the weight of these soon dropped to a mere 700g (1lb 8oz) and even less. This led to reduced fertility among much of the breeding stock and other health problems became apparent. Subsequently the norms regarding weight were altered to permit larger specimens, which has had a beneficial effect on both the fertility and life expectancy of these rabbits. In common with other dwarf breeds, Netherlands Dwarfs

Netherlands Dwarf with pink-eyes

Netherlands Dwarfs are very popular.

Netherlands Dwarf with blue eyes

Netherlands Dwarfs with Blanc de Hotot markings and a pink-eyed albino.

remain more difficult to breed than larger breeds. The litters are small: on average consisting of just two offspring, but this does not reduce the enormous popularity for these rabbits. There is not a single country that has rabbit enthusiasts where this breed is not to be found.

Netherlands Dwarf with blue eyes originate from Germany where they were bred by Lohse of Dippolswalde and Kluge of Hohndorf. Their new strain was developed using pink-eyed Netherlands Dwarf, non pedigree stock, Dutch, and Vienna Whites. These rabbits were first shown at the World Show in Leipzig in 1919. These rabbits were of somewhat coarser build than the pink-eyed Netherlands Dwarf rabbits but through careful selection and continued cross-breeding with pink-eyed Netherlands Dwarf stock, the blue-eyed strain is now more or less identical to the other Netherlands Dwarf rabbits.

Both types of Netherlands Dwarf were discovered by British enthusiasts during the 1950s. Because similar rabbits had been bred in Britain for years under the breed name Polish, the British breeders named the new rabbits, which were imported from The Netherlands, Netherlands Dwarfs.

The Americans also discovered these white rabbits that were developed in Germany and in 1969 some were imported by American breeders together with the coloured versions that had subsequently been developed. The Netherlands Dwarf has since become one of the most popular breeds of rabbit in the USA. Although continental European rabbit fanciers differentiate between the coloured and white varieties of these rabbits, in Britain and USA no division is made between the pink-eyed and blue-eyed Dwarf "Polish" rabbits. Both are known as Netherlands Dwarfs.

Properties

There is a slight difference in character between the blue-eyed and pink-eyed Netherlands Dwarfs: those with blue eyes are somewhat more lively. Netherlands Dwarfs are one of the most popular breeds of rabbit in the world, with the pink-eyed albino being especially widely kept. Because of their small size, these rabbits can be kept in small hutches and are ideal for people with limited space. The breed is not just popular with serious enthusiasts, many are adored as household pets.

External characteristics

Netherlands Dwarfs are noticeably stocky in their build with a very short neck. The legs are neat and straight with small feet. The head is rounded, with a broad forehead and full cheeks. Seen from the side, the nose is strongly curved.

The eyes are fairly large and rounded and the ears are small and slim, close together and about 50mm (2in) long. There is no international standard for the weight of these rabbits. The range is from 1–1.5kg (1lb 3oz–3lb 5oz) with the standard in Britain and USA being in the middle of this range.

Coat

The coat consists of short, shiny hair with a substantial undercoat, and it feels soft to the touch.

Colour

The white varieties of Netherlands Dwarfs are pure white, without any pigmentation, and both pink-eyed and blue eyed types exist.

Special remarks

Brown-eyed Netherlands Dwarfs were developed in 1928 by a German breeder through cross-breeding with Blanc de Hotot rabbits but no examples appear to remain from this variety.

Albino Netherlands Dwarf

Blue-eyed Netherlands Dwarf buck

Netherlands Dwarf
(Coloured)

Origins

The coloured varieties of Netherlands Dwarfs were developed in The Netherlands in the 1930s from white Polish and a selected strain of wild rabbits. The most important breeders in establishing this new breed are generally regarded as Messrs Andrea (who developed the Thrianta), J. A. Schippers, Hoefman, C. W. Calcar, but above all J. Meyering from Opheusden in the Dutch Betuwe region.

The first of these to show the rabbits was Hoefman of Brielle, in 1938. The breed was recognised by the Dutch Rabbit Breeders' Association in 1940, and although the standard established at that time permitted all known colours, the first Netherlands Dwarfs were almost exclusively rabbit grey. The maximum weight at that time was 1.5kg (3lb 6oz) and the maximum length of ear was set at 70mm (2_in). The self black variety was developed after World War II. This was followed by the breeders developing steel grey, and various sable shades and then through cross-breeding a wide assortment of colours was developed.

Black otter Netherlands Dwarf

Black otter Netherlands Dwarf

Black and tan Netherlands Dwarf

The first Netherlands Dwarf with a specific colour pattern was a colour marked example but the breed now includes virtually every type of marking and colour marking found in other breeds.
These rabbits arrived in Britain during the later 1940s where they quickly established themselves under the name Netherlands Dwarfs. American enthusiasts did not

Rabbit grey Netherlands Dwarf

Medium blue sable Netherlands Dwarf

Steel grey Netherlands Dwarf

colours quickly won it many enthusiasts. Although these coloured examples are judged as a separate breed in continental Europe, the white and coloured examples are all regarded as Netherlands Dwarfs in the United Kingdom and United States.

Properties

Netherlands Dwarf rabbits are lively creatures that with their white brothers and sisters make up the most popular breed of rabbit in the world.

Their popularity is partially explained by their compact size which means that they can be kept by breeders who are restricted for space. In addition to this, virtually every colour and type of marking found on other breeds of rabbit is to be found with Netherlands Dwarfs.

discover them until much later, with the first being imported into USA in 1969, where they used the same English name. The breed's small size and wide range of

This ensures that there is a Netherlands Dwarf to suit almost every taste. The huge

following among serious breeders and fanciers is mirrored by the breed's popularity as a household pet.

External appearance

The coloured Netherlands Dwarfs have a very stocky body with extremely short neck. The legs are slim and straight with small feet, the head is rounded, with broad forehead and full cheeks.

Seen in profile, the nose is very curved and the ears are small and slim, standing together and upright on the head. They are about 50mm (2in) long. There is no international standard for the weight of these rabbits. In the Netherlands, the minimum weight is about 800g (1lb 12oz) in Britain and USA these rabbits may weigh around 1kg (1lb 3oz) and elsewhere in Europe they can be up to 1.5kg (3lb 5oz).

Coat

The coat of the coloured Netherlands Dwarfs is short, with a strong sheen, and feels soft to the touch.

Colours of the Netherlands Dwarf

In addition to the two white varieties, the Netherlands Dwarf rabbit is bred in virtually every colour, colour pointing, and marking that exists in rabbits.

Agouti

Agouti colours have different bands of colour on each hair. The undercolour is next to the body, which is grey with many rabbits. The tips of the hairs have colour bands which are known as ticking and between these two is the base colour.

Other characteristics of agouti colouring are a light coloured belly, insides and backs of the legs, underside of the tail, and lighter-coloured rings around the eyes.

Agouti colours include "hare" colouring (or a reddish brown known as russet (reddish base coat with black ticking and brown eyes), rabbit grey (light grey/brown with

Red Netherlands Dwarf

ticking and blue eyes), steel grey (light grey with black ticking and brown eyes), chinchilla (silver/white with black ticking and brown eyes), and blue dun (light grey with blue ticking and blue eyes).

Yellow and orange examples are in reality agouti coats without any ticking: they do have the light-coloured rings around the eyes that do not occur with self-coloured rabbits. These latter two colours both have brown eyes.

Self coloured

Plain coloured rabbits are referred to as self coloured. Such examples have no ticking or lighter patches under their tail, belly, or behind their legs.

black ticking and brown eyes), grey/brown (light brown with chocolate ticking), brown dun (light grey with chocolate ticking), blue/grey (light grey/brown with blue/grey

The body is one colour from head to tail, with the same colour on the belly, although this may be more matt than elsewhere on the

Blue Netherlands Dwarf

Black silver fox coloured Netherlands Dwarf

body. Self colours that occur with Netherlands Dwarf include blue, with blue eyes, black and brown, both of which have brown eyes.

Colour-points and markings

Coloured Netherlands Dwarfs are bred in almost every known form of marking and colour-point that exists in rabbits. There are two types of colour marking with rabbits: one that can be bred true from generation to generation, and the other which is far less certain, so that luck as well as understanding of genetics plays its part.

COLOUR-POINTS
This type of marking is similar to that found on Himalayan rabbits. The actual colour – which with these rabbits is almost always black – only appears on the extremities of the body (such as ears, nose, legs, and tail). The colour only occurs on the colder parts of the body where less blood circulates. These rabbits are born white or mouse grey with the dark markings only becoming manifest after a number of weeks. The eyes of this variety are always pink.

SABLE
The various sable colours (mainly blue and sepia) are a fairly common variety with Netherlands Dwarfs.

The sable colours can be dark, medium, or light. The sepia types have brown eyes while blue sable rabbits have blue/grey eyes. This type also has a fiery glow in its eyes when viewed from a certain angle.

TAN
Netherlands Dwarfs with tan markings are extremely popular. The markings are similar to those of a Tan rabbit.

The orange/brown of the tan can be seen around the eyes, on the front and insides of the jaw, on the chest and belly, and on the

Blue silver fox coloured Netherlands Dwarf

Amber coloured Netherlands Dwarf

Chinchilla coloured Netherlands Dwarf.

coat after five to six weeks. All the hairs that replace shed hairs from now on have silver tips, which actually means the ends have no pigmentation.

The length of time this process takes is dependent on the rate at which the rabbit sheds its hair. Silvering comes in three intensities but Netherlands Dwarfs are virtually always the medium silvering, which means that about half the hairs have silver tips and half are normal. Judges prefer specimens that are uniformly silvered.

AMBER

Amber coloured Netherlands Dwarfs have hairs with a brown/black tip that gives a subtle tint to the coat.
The ears, chest, nose, hindquarters, legs, belly, and lower parts of the shoulders and flanks are darker than the rest of the body. This darker colouring is most pronounced on the nose, ears, and belly. If the hairs are blown to separate them, the much lighter colour beneath can be viewed. Amber Netherlands Dwarfs have dark brown eyes.

insides and rear of the hind legs. There are also small tan patches on the feet. A tan strip runs around the neck and , along the rim of the jaw, ending in a tan triangle behind the ears.

There are also points of tan on a number of longer hairs in an area from the belly to midway along the rabbit and also on the hindquarters. The other parts of the body are black or blue. Occasionally a rabbit may have brown or lilac main colour with tan markings.

SILVER FOX
This marking is identical to those with tan markings, except that what is tan is silver/white with these rabbits. This type of rabbit is known in black, blue, brown, and lilac, although the black and blue forms are the most popular.
The eyes of the black and brown types are brown, while those of lilac and blue ones with Silver Fox markings are blue.

SILVERED
The silvered Netherlands Dwarf is derived from the Silver. The offspring are born plain coloured but start to get silvering in their

Himalayan marked Netherlands Dwarf

Rabbit grey Netherlands Dwarf at various stages of development

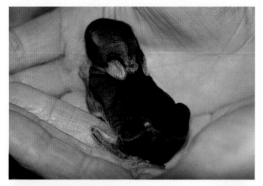

GREY/YELLOW

This colour is a diluted version of the amber colouring in which the colour marking is the same but the base colour is a far lighter yellow tinged with a blue haze instead of brown/black. The eyes are blue.

Netherlands Dwarfs with markings

Unlike the previous colour markings, various patterns, patches, and spots are more difficult to recreate with any certainty. Pairing two perfectly marked Netherlands Dwarfs together is no guarantee that their offspring will have similar markings, making breeding a hit and miss affair.

BLANC DE HOTOT MARKINGS

Markings like those of Blanc de Hotot rabbits are perhaps the most popular form of marking for Netherlands Dwarfs. Rabbits with these markings are white with black eye

Netherlands Dwarf with Blanc de Hotot markings

rings that are about 3–5mm (^{1}/$_8$–^{3}/$_{16}$in) wide. The rings should circle the eye as uniformly as possible. The eyes are always brown.

PAPILLON MARKINGS

Netherlands Dwarfs with Papillon (or English Butterfly) style markings are a recent arrival on the scene and they are not yet recognised in every country.

The body and head markings are similar to those on both Giant Papillons and English/American Checkered/Papillon/English Butterfly rabbits. These markings include the butterfly, eye rings, thorn, cheek flashes, and coloured ears.

The butterfly is a dark patch on the nose which should be sharply defined and run from one corner of the mouth to the other. The mandatory circles around the eyes should be of equal thickness. The thorn is a marking in the centre above the nose and fully coloured markings of the cheeks are also a required feature.

The body should have a coloured dorsal stripe from behind the ears to the tail which judges prefer to see of even thickness and sharply defined.

Finally these rabbits should have small spots on their sides, ideally round, which should be well separated. This type of marking can theoretically be produced in any colour but to date they are mainly marked in black.

HARLEQUIN MARKINGS

Netherlands Dwarfs with harlequin markings should be the same colours and patterns as Harlequin rabbits.

This ideally means rabbits with heads of two colours that are split precisely in the centre, with which the chest beneath the black part is orange with a black leg, and beneath the orange half the chest is black with an orange leg.

The ear of the orange side of the head should be black and vice versa. The body should be

Breeds with unusual types of coat

Rex

Origins and history

Mutations (or sports as they are also called) occur quite frequently. The reality is that many breeds of rabbit came into being through a mutation which was then fixed through a breeding programme and selection.

Rex rabbits are no exception. The first rabbit with rex hair that breeders knew about was born at the village of Louché-Pringé in

Fine castor Rex buck

banded with black and orange zebra stripes. The numbers are not important but they must be clearly defined.

DUTCH MARKINGS
Dutch markings mean two coloured plates on the head that include the eyes, cheeks, and ears but do not extend to the nose or neck.

The correct markings form an inverted V blaze on the forehead. This type of rabbit is coloured from the middle of its back to its tail with a sharp division between the coloured and white parts of the body that is also very straight. The feet on the hind legs are always white.

Dalmatian Rex

Young Rex rabbit

Albino Rex rabbits are very popular.

France in 1919. This rabbit was the colour now known as castor. The village priest, named Gillet, was impressed by the creature and decided to breed it. Several of these Rex rabbits (castor coloured) were first displayed at a major international show in Paris in 1924. This was the breakthrough for this new breed that quickly found great interest among breeders.

Unfortunately, in addition to breeders, there were others whose main interest was the absurdly high prices that were paid for a Rex pelt so that the breed quickly degenerated because the does were commercially over bred too frequently. This setback caused the fur trade to lose interest because the coats were no longer of adequate quality.

Fortunately a few serious breeders remained loyal to the breed and did everything possible to restore the Rex back to its former glory.

Fortunately this succeeded, which is obvious, since today's Rex is a healthy and feisty rabbit with a coat that is generally of excellent quality.

Two young blue Rex rabbits

Albino Rex buck

Every Dalmatian Rex has a unique pattern of spots.

Young yellow Rex

By introducing other breeds and colours, Rex rabbits can be admired in numerous colours and markings. The original castor colour is still the most popular in the majority of countries. The castor Rex from France was not the only mutation that attracted attention. There were subsequently regular appearances of rabbits with a similar type of coat.

This happened in Lübeck in Germany in 1924, in Chartres in France in 1927, and in Schoonhoven in The Netherlands in 1938. Attempts to cross breed the various lines of Rex, which were not related to each other, did not produce any offspring with the same type of coat. This proved that there are different mutations that cause this type of coat.

Properties

The extraordinarily velvety soft hair of a Rex rabbits coat, combined with the elegance, spirited but good natured character of this French breed deserves greater appreciation as a household pet. Of course some of them are kept as pets but the majority of fanciers are still breeders.

The Rex is one of the most popular breeds across the world with many breeders busy showing them and developing ever new colours.

External characteristics

Rex rabbits have a slightly long and lean form of body with legs of normal length that seem longer because their hair is shorter than most other breeds.

Viewed in profile, the head appears quite arched. The bucks have particularly imposing heads with substantial cheeks. The average length of ear is 120mm (4¾) and the weight is 3–4kg (6lb 10oz–8lb 13oz). Some countries judge different colours separately and have different weight standards for different colours.

Coat

The quality of the coat is the most important feature of the Rex rabbit. The hairs are quite short at about 15mm ($^5/_8$in) and stand up straight.

This plus the fact that the hairs are the same length across the whole body, give the coat an exceptional velvety appearance.

The hairs are fine but tough and they are densely formed with a substantial amount of undercoat which helps the coat's density. The whiskers of Rex rabbits are always curly.

Colours

Rex rabbits are bred in almost countless colours and markings with new ones being

Lilac English Rex

Castor Rex doe

Head of a Dalmatian Rex doe

continually added through cross-breeding with other breeds. Some of the well-known and lesser known colours are described below.

AGOUTIR

Agouti coloured rabbits have different bands of colour on each hair. The colour nearest the body, or undercolour is usually grey or blue. The hair on the tips of the hairs is known as ticking and the colour in between is the base coat.

The best way to see the transition between the three colours is to blow against the hairs to separate them. Other characteristics of agouti colourings are lighter colour on the belly, light rings around the eyes, and lighter areas on the insides and backs of the legs, and underneath the tail.

The agouti colours include rabbit grey (light grey/brown base coat with black ticking and brown eyes), chinchilla (silver/white with black ticking and brown eyes), German Lux rabbit colouring (orange with silvered pale

blue ticking and blue eyes), opal (red/brown with lilac ticking and brown eyes), and castor (mahogany with black ticking and brown eyes).

Yellow and orange are in reality agouti colours without any ticking: these colours have light rings around the eyes which are not seen with self-coloured rabbits. Yellow and orange rabbits have brown eyes.

SELF COLOURED

Plain coloured rabbits are known as self-coloured. These rabbits have no ticking in their coat or any lighter patches beneath the tail, on their belly, or on the legs.

The coat is the same colour from head to tail and the hairs should be the same colour right down to the roots. Rex rabbits are bred in self black (with brown eyes), blue (with blue eyes), chocolate brown (with brown eyes), and lilac (with blue eyes).

Plain white Rex rabbits, known as Ermines, are extremely popular; these are recognised as albinos (with pink eyes) and also with blue eyes.

COLOUR-POINT MARKINGS

Numerous different types of colour markings that can be true bred are available with Rex rabbits. Two rabbits with these markings will produce offspring with similar colour markings, unlike the experiences with other types of markings which are not genetically inherited.

The colour markings known for Rex rabbits include amber colouring, Silver Fox and tan (in various colours), Himalayan, and various sables.

MARKINGS

Other types of markings, such as patterns or spots do not breed true so that breeding two perfectly marked specimens is no guarantee of correctly marked offspring, so that breeding rabbits with markings is not a simple matter.

An extremely attractive and popular type is the Dalmatian Rex, which originates from Switzerland, which is bred with black, brown, or blue spots, but also with a combination of colours. Rex rabbits with Dutch markings are extremely rare.

Special remarks

There are two other types of Rex rabbit that originated in Britain and are almost exclusively bred and kept in Britain. The first is the Opossum Rex that was bred in 1924 by T. Leaver of Kent. These rabbits have fairly long hairs of about 25mm (1in) that stand upright from the body but curl strongly at the ends.

The striking feature is the silvering of the ends of the hairs all over the body with the exception of the head, legs, and ears. These rabbits are still bred on a very small scale in Britain. The other type of Rex is the Astrex which has hair that curls tightly against the body. These rabbits were developed in the 1930s. The head, ears, and legs of the Astrex are of more or less normal hair. This breed has almost disappeared.

Litter of Satin rabbits

Satin

Origins

Satin rabbits get their name from the striking satin sheen on their coats. This occurred originally in a litter of Havana rabbits belonging to the American breeder Walter Huey of Pendleton, Kentucky. These rabbits, that were born in 1932, created quite a stir when they were shown to the public at Louisville in 1934.

Initially the intention was to introduce Satin rabbits as a new fur breed but interest from fur breeders was limited. Another American breeder named Price, from Phoenix, Arizona, wanted to try to improve the mutated gene by cross-breeding with New Zealand Whites, which were then the most popular meat producing rabbit in the United States. These crossings eventually led in 1938 to ivory coloured Satins that were genetically albino rabbits but which had a yellow tinge due to the unusual texture of their coats. Price never achieved his objective so that the Satin remained a fancy rabbit with no value for the meat and fur trade.

Despite this lack of commercial value, some breeders remained loyal to the breed which is now a much loved breed for showing. The first Satin rabbits were exported to other

Orange Satins

Ivory Satin

Head of an orange Satin

Satin

skull. The head is especially well developed with bucks. The ears are sturdy, with an average length of 110mm (4¼in).

Coat

The Satin rabbit's name is derived from its exceptional coat with its unusual texture, which is genetically recessive by comparison with normal coats.
The hairs are fine and give the coat an exceptional deep sheen that is only found on this breed. Judges at shows pay particular attention to the density of the coat. Satin rabbits have more hair in their coat than any other type of rabbit.

Colours

Satin rabbits are bred in more colours than can be dealt with here but the most popular

Albino Swiss Fox rabbit

Blue-eyed white Swiss Fox rabbit

countries from the USA in 1947. The ivory coloured Satins arrived in the United Kingdom that same year where breeders soon established a range of additional colours by cross-breeding the Satins with Rex rabbits. The influence of British breeders resulted in Satins now being a much appreciated breed at shows throughout Europe.

Properties

Satin rabbits have a special appearance and they are generally calm with a gentle nature. These rabbits are certainly ideal pets for children but the majority of this breed are kept by serious breeders who compete in shows. The breed is not widely known by the public.

External appearance

Satins weigh 2.5–4kg (5lb 8oz–8lb 13oz) and have bodies of a normal length that are well rounded. These rabbits have a large head with well developed cheeks and a broad

one continues to be the original ivory. In principle, any colour or marking can be achieved with Satin rabbits by crossing them with other breeds.

The first generation offspring do not have satin coats although they bear the satin factor gene. There is a reasonable chance that satin haired offspring will be produced only when two rabbits bearing the gene are bred together. The most widespread colour is the ivory – which is an albino. There are also white Satins with blue eyes but these are quite rare.

Self coloured Satins include black and an original chocolate brown colouring with brown eyes, and blue and lilac with blue eyes. Other colours include agoutis such as reddish brown (russet base coat with black ticking and brown eyes), and chinchilla (silver/white with black ticking and brown eyes).

Orange Satins are in reality an agouti colouring without any ticking; with light rings around the eyes, a light belly, and lighter underside of the tail plus lighter in-

sides and backs of the legs. Finally, there are also Satin rabbits with true bred colour markings such as Himalayan, and sable, although these are rare outside the United Kingdom

Swiss Fox

Origins

The blood flowing through the veins of Swiss Fox rabbits that we see these days at shows includes Havana, Chinchilla, and Angora. One of the most important breeders and enthusiasts for this breed was a Mr Müller of Switzerland, who cross-bred Havana and Angora rabbits in the 1920s. His new "creation" was recognised in 1925 under the name Schweizer Fuchskaninchen.

The original purpose of Mr Müller was not the development of a show species but to produce a pelt similar to fox fur, which was

Blue Swiss Fox rabbit

after Müller, German breeder Leifer of Leipzig managed to develop a similar breed. His first specimens were shown as Blue Fox rabbits in Leipzig in 1932.

There were other German breeders who independently of Leifer sought the same objective but their results were less successful than his. Swiss Fox rabbits were imported in some numbers into German and found their way to other countries as well. They were recognised in The Netherlands in 1933 when they were also shown in Britain for the first time.

Despite this, it took until the 1980s before the breed was officially recognised in Britain. In common with many other breeds that were purely kept for pleasure and competition, there were so few Swiss Fox rabbits of any importance in existence at the end of World War II in the other countries of Europe.

However, enthusiasts were able to find new stock in Switzerland which had remained neutral throughout the war. Since then, the Swiss Fox has become a well-known sight at

Young yellow Angora (see pp 306–309)

widely worn in those days. Neither Müller or the other breeders of the Swiss Fox succeeded in this ambition. The texture and hardness of the coat is quite different to that of a fox and quite different to other rabbits as well.

This Swiss breed did become very popular though with rabbit enthusiasts. Several years

Blue-eyed white Swiss Fox

Albino Angora rabbit (see pp 306–309)

Young black Angora (see pp 306–309)

the fur trade but due to lack of interest for the pelts, the breed remained in the hands of rabbit fanciers who bred these rabbits for their looks. The breed does not have the wide following of breeds such as the Flemish Giant or Dutch rabbits but has a small band of enthusiasts among those who show rabbits.

The breed can be seen regularly at shows but is virtually unknown as a household pet which is a shame because it is ideal for those who like a rabbit with a longer coat but not the trouble involved with grooming the coat of other long-haired breeds. In common with any other rabbit, Swiss Fox do need some grooming from time to time but they need significantly less care and attention than Angora rabbits.

External characteristics

Swiss Fox rabbits have muscular bodies with quite striking heads that are well developed with substantial cheeks. There is no neck to speak of. In common with most rabbits, the buck's head is more pronounced than a doe's. The average length of the ears is 110mm ($4^5/_{16}$in). Swiss Fox rabbits weigh 2.5–4kg (5lb 8oz–8lb 13oz).

Coat

Judges pay close attention at shows to the fine texture and density of the hairs in the Swiss Fox rabbit's coat. The hairs should not be woolly and are preferred to be straight,

Newly clipped albino Angora (see pp 306–309)

most shows, albeit that the breed is not to be found everywhere.

Properties

Swiss Fox rabbits have a calm temperament. These rabbits were originally intended for

Angora rabbits need plenty of attention to their coats.

without any curls. The copious undercoat pushes the harder guard hairs upwards so that they do not lay sleek.

The hairs are 50–70mm (2–2½in) long except on the head, belly, and legs where they are much shorter.

Colours

Swiss Fox rabbits are bred in various colours but the white examples, with pink eyes or blue ones, are the most popular.

There are also self-coloured black, blue, and havana Swiss Foxes and also the colour known as blue fox. Finally, there are also chinchilla and yellow Swiss Foxes but these are quite rare and not recognised in every country.

Angora

Origins

Angora rabbits with their full coat cannot cope with too much warmth or damp con-

Blue Angora doe with her young

Blue Angora

Brown Angora

ditions. These rabbits cannot be washed because the moisture does not dry out readily and makes the coat more susceptible to tangles. The hair needs brushing every day to prevent tangles that cause the rabbit problems and can hamper its freedom of movement.

The grooming needs to be carried out very carefully because over-enthusiastic brushing can cause loss of hair. From time to time, depending on the length of hair and its density, the hair will need clipping. This is usually about once every three months. If this is not done, or not carried out properly, the Angora's skin can suffocate, causing death in extreme cases.

Angora rabbits with a full coat cannot be kept on a bed of straw, hay, or sawdust, because this will become trapped in the hairs. Angora rabbits that are regularly shown are kept in hutches with mesh floors. Clipped Angora rabbits, however, can be kept on straw without virtually any problems provided the loose straw pieces are removed from the coat daily.

Breeding of Angora rabbits has its own special difficulties because of the length of the coat. The covering by the buck may not always be successful due to the length and

Angora rabbits are born dark but become lighter.

Head of a brown Angora doe

hairs. Angora rabbits that are regularly shown are kept in hutches with mesh floors. Clipped Angora rabbits, however, can be kept on straw without virtually any problems provided the loose straw pieces are removed from the coat daily.

Breeding of Angora rabbits has its own special difficulties because of the length of the coat. The covering by the buck may not always be successful due to the length and texture of the coat and does may experience problems giving birth and suckling their young unless the surplus hair is removed.

This breed's origins are as producers of wool. An Angora rabbit can be shorn or clipped for the first time at six to eight weeks. The coat grows about 80mm ($3^1/_8$in) in three months.

Depending on the rate of growth and density of the coat, an Angora rabbit can produce 1kg (2lb 3oz) of wool or more each year.

texture of the coat and does may experience problems giving birth and suckling their young unless the surplus hair is removed.

This breed's origins are as producers of wool. An Angora rabbit can be shorn or clipped for the first time at six to eight weeks. The coat grows about 80mm ($3^1/_8$in) in three months.

Depending on the rate of growth and density of the coat, an Angora rabbit can produce 1kg (2lb 3oz) of wool or more each year.

Properties

Angora rabbits with their full coat cannot cope with too much warmth or damp conditions. These rabbits cannot be washed because the moisture does not dry out readily and makes the coat more susceptible to tangles. The hair needs brushing every day to prevent tangles that cause the rabbit problems and can hamper its freedom of movement.

The grooming needs to be carried out very carefully because over-enthusiastic brushing can cause loss of hair. From time to time, depending on the length of hair and its density, the hair will need clipping.

This is usually about once every three months. If this is not done, or not carried out properly, the Angora's skin can suffocate, causing death in extreme cases.
Angora rabbits with a full coat cannot be kept on a bed of straw, hay, or sawdust, because this will become trapped in the

Clipping an Angora requires skill.

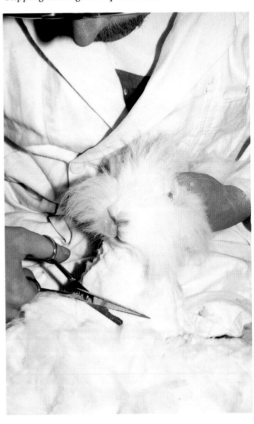

External characteristics

Angora rabbits weigh 3–4kg (6lb 10oz–8lb 13oz). They are fairly stocky animals and with the length of hair, their necks are completely unseen.

Their legs are of normal length but should be sturdy and muscular. Angoras have a broad head that should be fairly short with strong, thick ears that stand erect of about 120mm (4¾in) long.

Coat

Judges pay special attention at shows to the quality, length, texture, and density of the coat. An Angora in good condition should have a coat at least 60mm (2⅜in) long in order to impress but judges prefer even longer coats.

In common with other rabbits, Angoras too have different types of hairs: the undercoat or wool, base coat and outer hairs. The soft wool must predominate in an Angora's coat. Plumes on the ears are essential for show success.

Colours

Angora rabbits are bred in a diverse range of colours. The best known colour is albino (white with pink eyes) but white with blue eyed Angoras are also recognised. In addition, these rabbits can be seen in black, brown, yellow, and blue plus sable colours, although these are rare and not recognised in every country.

With all adult Angora rabbits, the actual colour of their coat can only be fully seen on the short hairs around their nose and on the insides of their ears.

The remainder of the body has a coat that is much lighter in tone.

Special remarks

There are various different types of Angora rabbit.
French Angoras have a different type of hair than the other types of Angora, with harder hairs and a different moulting cycle. French Angoras are plucked rather than shorn or clipped.

Young yellow Angora rabbit

Rabbit grey Mini Rex

Dwarf breeds with unusual types of coat

Mini Rex

Origins

Mini Rex rabbits are the result of cross-breeding Netherlands Dwarfs with Rex rabbits. The first examples were white and sable. These rabbits were first exhibited as Netherlands Dwarfs with Rex coats at the 1966 Dutch Ornithophilia show by the Boydon breeders group from Den Dolder in The Netherlands. The breed was recognised under the Mini Rex name in Britain in 1990 where the breed standard for build is similar to that of the Rex, except in the matter of

Black Dwarf Angora

overall size. The breed is also known in the USA where is has many enthusiasts.

Properties

The character and other characteristics of this rabbit are similar with those of the

Dwarf Angora (see pp 312–314)

Dwarf Angoras are ideal pets for those with less space.

Netherlands Dwarfs. Mini Rex rabbits (or Netherlands Dwarfs with Rex coats) are bred in virtually every rabbit loving country but are not uniformly popular. The combination of their lively nature and attraction of stroking them makes them popular pets as well as show specimens.

Dwarf Angora

External characteristics

Mini Rex rabbits have the same build as Netherlands Dwarfs.

Coat

The quality of the coat is the most important point of a Mini Rex rabbit. The hairs stand out from the body and are the same length across the entire body to give an effect like velvet. The hairs are fine yet tough and very densely formed. Mini Rex rabbits have a substantial amount of woolly undercoat, which aids the coat's density. The whiskers are always curly.

Colours

Mini Rex rabbits are theoretically bred in every colour that is produced with Netherlands Dwarfs. Since the breed is relatively new, many colours, markings and colour markings are still being developed.

Angoradwerg

Origins

Dwarf Angora rabbits came into existence at about the same time in a number of countries through the cross-breeding of

Albino Dwarf Swiss Fox rabbit

Angora rabbits with other breeds. The Belgian creator was a breeder named Born of Liège who crossed small Angoras with albino Dwarf Polish rabbits.

The new breed he developed was recognised in Belgium in the late 1980s and was also recognised in France, but despite much effort by breeders is not yet accepted in Germany or The Netherlands. Interest in the breed is growing every day, so that recognition must only be a matter of time.

Dwarf Angora

Properties

This breed is ideal for those who wish to have a long-haired rabbit but with too little time to look after a full-sized Angora. The Dwarf Angora does though still need careful attention for its coat.

With their full coat, these rabbits cannot withstand too much warmth or damp conditions. The coat cannot be washed because it is difficult to dry and liable to tangle. It

Two Dwarf Swiss Fox rabbits (see p314)

will be necessary to brush your Dwarf Angora every day in order to keep the hairs from becoming tangled. Check each day for any tangles but prevention is better than cure. Brush or comb carefully because the coat can be easily damaged.

Depending on the rate at which the hair grows, your rabbit will need clipping about every three months. This is essential to allow the skin to breathe.

A Dwarf Angora rabbit cannot be kept on a base of straw because it will get caught up in the soft hairs. Dwarf Angoras that have been clipped can be kept on straw provided that loose pieces of straw are removed daily from the coat. Most people who show Dwarf Angora rabbits keep them on a mesh floor. Before a Dwarf Angora is mated, she will need to be clipped.

Without doing so, covering her will be difficult and she will have problems with suckling her young.

External characteristics

The Dwarf Angora should look just like a smaller version of the Angora rabbit. These rabbits are not true dwarf rabbits despite their name.

At the present time the average weight for them is about 1.75kg (3lb 13oz) with an ideal weight of 1.5kg (3lb 5oz).

Coat

The coat of a Dwarf Angora has the same texture and density as that of a normal Angora rabbit, although the hair is shorter. The minimum standard for show specimens is 50mm (2in).

Colours

At present most Dwarf Angoras are white examples (both albino with pink eyes and white with blue eyes). Havana brown and black examples with brown eyes are beginning to appear more often at shows.

Dwarf Swiss Fox

Origins

The Dwarf Swiss Fox is bred in Germany and The Netherlands by crossing Swiss Fox rabbits with Dwarf Polish. The breed was recognised in The Netherlands in 1994.

External characteristics

This rabbit has a similar build to Netherlands Dwarfs but should have a coat like that of the Swiss Fox, although shorter. This is a dwarf breed with a short, stocky body and rounded small head.

The coat is of medium length, should feel soft to the touch but neither woolly or straight.

These rabbits are mainly available in white (both albino with pink eyes and with blue eyes).

Albino Dwarf Swiss Fox

Addresses

British Rabbit Council
Purefoy House,7 Kirkgate
Newark NG24 1AD
United Kingdom

British House Rabbit Association
PO Box 346
Newcastle-upon-Tyne NE 99 1FA
United Kingdom

The National Gerbil Society
373 Lynmouth Avenue
Morden
Surrey SM4 4RY
United Kingdom

The National Fancy Rat Society
14 Clayhall House
Somers Close
Reigate
Surrey RH2 9ED
United Kingdom

National Mouse Club
22 Malham Road
Rastrick
Brighouse
W. YorkshireHD6 3JY

National Hamster Council
PO Box 154
Rotherham
South Yorkshire S66 0FL

British Hamster Association
PO Box 825
Sheffield
S17 3RU

National Cavy Club
Olney Park Cottage
Yardley Road
Olney
Bucks HP13 5NN

Black Netherlands Dwarf Rex

Orange Satin

Acknowledgements

The publisher and author thank the follow persons for making their attractive animals available and for their help in making the photographs possible for this book.

H. Akkerman, Fr. Aleven, L.A. van Bakel, Comb van Beek, C. van de Berg, T. Blokker, C. Bregman, A.W. Cannell, the Captijn family, H. Cornelissen, L. van Dalen. J Drury, C. Duyvekam, A.E. Derksen, M. Derksen, A.M. van Dongen, L. Donkersteeg, H. Dutewaert, C. van Empel, L. Everitt, H. Fennis, H. van de Geest, F. Gidding, D. Goudriaan, the Goedhart-Pappot team, D. Grau, H. Hak, the Harks family, F. den Hartog, A. Hendriks, A. van Hinthum, the den Hollander family, N.G. Hoornsman, J.C. Hulleman, J.W. Jansen, D. de Jong. H. Jonker, the Kamps team, J. Kanen. the van Kapel, J. van Kessel, the Kogels familiy, H. Kool, M.J.A. van Kooten, Mr. Kraayenveld. the van Kruistum team, W. Kreydt, T. Kwetters, W. van Laar, D. van Leeuwen, W. van Leeuwen, H. Lindeboom, W. Lindeboom, Judith Lissenberg, E. van Manen, the Meeldijk family; P. Megit, Ben Mimpen, W. Monshouwer, D. van Muijden, M.J. van de Muijden, Yolanda van Mul, G. Niesing, the van Nuland family, J. van Oirschot, the "Onze Sport" team, A.G. Oomen, Joyce den Otter, W. den Otter, the v.d. Pas family, J. van de Pavert, the Pennings family, H.G. Philipse, G. Prins, Gaby Prust, Rabbiminimus, G. Reijersen, the RodeToren team, the van Roelofs-Schmit team, J. van Rooij Sr., J. van Rooij Jr., L.A.G. van Rooij, A. van Rooiien, the van Rumpt family, A. Salari, the Satijn team, R. Schellevis, G.J. Schotman, J.J. Schreuder, K. Seepers, M. van Setten-Blok, J. and M. Stilkenboom, J.G. Tres-v.d. Linden, the de Veer family, P. v.d. Ven, H. van Venrooij, M. Verhoeven, J.F. Verrips, L. Vervecken, J.H. Verweij, Cees Vos, G. de Wit, J.G.P. Wolberts, W. van Zuilichem, H. Ziel and O. Zumbrink.

Special thanks are due to two people who are both experts in their own fields. These individuals were always to give help when it was necessary . Mr. A.G. Oomen (an official judge) helped with the chapter on rabbits while Mrs. Judith Lissenberg (chairperson of NMC) was a great help with the chapters about rodents. Their input and years of experience have benefited this encyclopedia. Thanks are due too to Mrs. L. Donkersteeg en Mr. B. Mimpen, who gave such generous help. Special thanks go to the committee and members of V.K.P.V. of Veenendaal, who were kind enough to arrange a place for the animals to be photographed and for their tremendous help.

Thanks are also due to Fur and Feather, the official publication of the British Rabbit Council for their help in providing photographs of a number of British breeds of rabbit and also to Mr. A. Vandenbroecke of the Speciaalclub Het Belgisch Raskonijn for information about a number of Belgian breeds of rabbit. Finally, thanks are due to Drs. F. Petrij for information about dwarf hamsters.

Two weeks old Alaska rabbits

Magpie doe

Photo acknowledgements

Almost all the illustrations were photographed by the author. Additional material was provided by Mrs. Judith Lissenberg of The Netherlands, XXX of Belgium, and Fur & Feather of the United Kingdom.

Judith Lissenberg provided approx. 80 photographs

Debbie Duocommun (the Rat Lady), 2 photographs

Mary-Ann Isaksen 2 photographs

Pat Gaskin (Fur & Feather), 13 photographs

Black and tan Belgian Hare

Steenkonijn or "Stone rabbit"

Index

Thüringer

English doe with dark amber markings

Californian

Perlfee

Havana

Havana

Steel grey Dutch rabbit

Rabbit grey Viennese

Brown Tan doe

Blue Papillon

Light black Silver

Sable

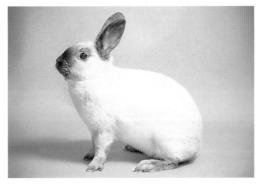